Tyre Retreading

Also of interest

Rubber Nanocomposites and Nanotextiles.
Perspectives in Automobile Technologies
Banerjee (Ed.), 2019
ISBN 978-3-11-064089-2, e-ISBN 978-3-11-064387-9

Rubber Analyis.
Characterisation, Failure Diagnosis and Reverse Engineering
Forrest, 2019
ISBN 978-3-11-064027-4, e-ISBN 978-3-11-064028-1

Rubber.
Science and Technology
Princi, 2019
ISBN 978-3-11-064031-1, e-ISBN 978-3-11-064032-8

Smart Rubbers.
Synthesis and Applications
Polgar, von Essen, Pucci, Picchioni, 2019
ISBN 978-3-11-063892-9, e-ISBN 978-3-11-063901-8

e-Polymers.
Editor-in-Chief: Seema Agarwal
ISSN 2197-4586
e-ISSN 1618-7229

Tyre Retreading

Edited by
Bireswar Banerjee

2nd Edition

DE GRUYTER

Author
Bireswar Banerjee
B – 12/3, Karunamoyee Estate
Salt Lake
Calcutta, 700091
India

ISBN 978-3-11-064090-8
e-ISBN (PDF) 978-3-11-064389-3
e-ISBN (EPUB) 978-3-11-064095-3

Library of Congress Control Number: 2019931192

Bibliographic information published by the Deutsche Nationalbibliothek
The Deutsche Nationalbibliothek lists this publication in the Deutsche Nationalbibliografie; detailed bibliographic data are available on the Internet at http://dnb.dnb.de.

Typesetting: Integra Software Services Pvt. Ltd
Printing and binding: CPI books GmbH, Leck
Cover image: RapidEye/E+/gettyimages

www.degruyter.com

Preface

Tyre retreading is an established technology in the recycling of worn tyre carcasses. Indeed, the term 'retread' was introduced during World War II by ex-army officers summoned up on duty.

About 85–88% of worn casings can be retreaded and deployed as fresh tyres, and if maintained properly, 80% of casings that have been retreaded once can be retreaded again and reused. An effective functional tyre is made by replacing the old tread with a new one by appropriate processing with suitable rubber compounds. A correctly retreated tyre is able to provide the same mileage, rolling resistance and other operational properties as a new tyre. The retreaded tyres are tested and their dependability and safety assessed according to the specifications for automobile tyres, truck tyres, off-the-road (OTR) tyres and aircraft tyres (taking into consideration, for example, the eventual extreme stresses experienced by aircraft tyres).

Because of the guaranteed and reliable performances of retreaded tyres but at a lower cost than new tyres, in 1993 the USA allowed retreaded tyres to be used on all government vehicles under a 'Federal Acquisition, Recycling and Waste Prevention' order. The foremost advantage of retreading is its cost effectiveness, with a retreaded truck tyre costing one-fifth that of a new truck tyre.

Tyre retreading is friendly to the environment as an alternative to buying a new tyre. A considerable amount of energy (equivalent to 84 litres of oil) is consumed in the manufacture of one new truck tyre, whereas 26.5 litres of oil is used to retread a tyre. In the USA, 15 million truck tyres are retreaded, thereby saving 100 million gallons of oil every year. In other words, retreading, via recycling and conservation, contributes to saving Earth's depleted resources.

Since there is no concise guide book on the manufacturing technology involved in tyre retreading, this volume of work attempts to fill this gap by describing the effective practical techniques of tyre retread manufacturing and other relevant topics of importance. This handbook comprises six chapters. In Chapter 1, details are provided regarding the selection of elastomers, the effective compounding ingredients, and the processing systems for retreading. Chapter 2 describes the processability and characterisation of the compounding ingredients used in retreading rubber compositions. Chapter 3 outlines the non-destructive testing of worn-out tyres, and elaborates on the processes of manufacturing retread in figures. Chapter 4 presents a comprehensive overview of 'green tyre' technology, achieved by substituting the common reinforcing agent carbon black with highly active silicas, as well as a discussion of other nanofillers used in tyre treads to achieve better wet grip and reduced rolling resistance.

Chapter 5 outlines up-and-coming trends in the use of elastomer nanocomposites, and emerging technologies that are likely to grow and move forward. It also reviews the 'Run-Flat Tyre', which is already on the tyre market for cars. Finally,

https://doi.org/10.1515/9783110643893-201

Chapter 6 covers the testing methodologies used in the analysis of polymers and the chemical analysis of compounding ingredients with relevance to quality assurance.

The main focus of this project was to offer a concise volume to the tyre retread manufacturer, technologist, student and researcher. The chapter contributors are all proficient scientists and academics in their respective fields of research, and their participation has guaranteed the scope of this handbook. I thank all the contributors for their efforts in preparing their chapters.

I express my great appreciation to Ms Helene Chavaroche, Commissioning Editor Publishing, for her endeavours to make this project a successful one. Pre-editing of the manuscript was done by Mrs Eleanor Garmson – the Development Editor, I express my gratitude to her.

I also express my sincere thanks to all the staff of Smithers Rapra who assisted in the preparation and publication of this handbook.

Bireswar Banerjee
2015

Contents

List of contributors

Samar Bandyopadhyay
Hari Shankar Singhania Elastomer & Tyre Research Institute (HASETRI)
P.O. Tyre Factory
Jaykaygram
Dist. Rajsamand
313324, Rajasthan
India

Bireswar Banerjee
B – 12/3, Karunamoyee Estate
Salt Lake
Calcutta, 700091
India

Tuhin Chatterjee
Rubber Technology Centre
Indian Institute of Technology
Kharagpur, 721302
India

Kinsuk Naskar
Rubber Technology Centre
Indian Institute of Technology
Kharagpur, 721302
India

Dipak Kumar Setua
Defence Materials and Stores Research & Development Establishment (DMSRDE)
DMSRDE – Post Office
G. T. Road
Kanpur, 208013
India

https://doi.org/10.1515/9783110643893-202

Bireswar Banerjee

1 Rubbers, compounding ingredients and their criterion in tyre retreading

1.1 Introduction

In the long history of automobile tyre manufacturing the following milestones are notable. In 1895, Michelin introduced its pneumatic automobile tyre. Previous to that, in 1888, Dunlop had patented the pneumatic tyre in the UK. Goodyear launched the first nylon-reinforced tyres in 1947 and, in the same year, BF Goodrich announced the tubeless tyre. Even more interestingly, in 1992 Michelin introduced the first durable 'green tyres', which were composed of rubber compounds reinforced with silica in place of the usual reinforcing agent carbon black.

In 1912, Marion Oliver patented a 10 ft and 12 ft long tread design, which was used in precure treads and marketed to other retreaders. In the 1980s, the tyre retreading business rapidly became globalised, and European companies began marketing equipment and manufacturing techniques in the USA. A new type of extruder was introduced to produce truck tyre retreading tread.

Also in the 1980s, a non-destructive method for the careful inspection of worn tyres was established. Then, in the 1990s, advancements began to be made in retreading using high-tech computerised devices. Technology for the primary inspection of worn tyres using an inflated inspection spreader and improved X-ray equipment has become available since then. Shearography, a new and improved off holography specially designed for the high-speed production and testing of tyres, can determine their viability for retreading. The inspection spreader device works by applying sufficient air pressure into a used tyre that any damaged sections of the casing become distorted, thereby allowing any defects to be identified easily.

The basic rubbers or elastomers used in compounding for tyre retreading purposes are high-molecular-weight (MW), amorphous, viscoelastic polymers with high elasticity. Rubber compounds prepared based on natural rubber (NR) with blends of synthetic elastomers such as polybutadiene rubber (PBR) and styrene-butadiene rubber (SBR) become soft and will flow during extrusion of the tread profile and moulding of the tread to give a flawless tread design with appropriate heat and graduations of stress during the moulding process.

https://doi.org/10.1515/9783110643893-001

For some special applications, tyre retread may require high weather, chemical, heat and oil resistance characteristics, which can be obtained by blending of NR, general purpose and special purpose synthetic rubbers [1].

The commonly used rubbers for tyre retreading applications are as follows:
- NR (isoprene)
- SBR
- PBR
- Chloroprene rubber (CR) for specific applications
- Ethylene propylene diene monomer (EPDM) rubber for special applications (in blends with NR)
- Reclaim rubber (as compounding ingredients)

An overview of the processing behaviour, physical/mechanical characteristics (such as hardness, tensile strength, modulus, abrasion, tear and flexing resistance), and chemical and oil resistance while operating in a particular environment is presented in this chapter. Technologists associated with the tyre retreading industry control rubber processing to determine the correct viscosity to function properly at a particular processing temperature, the rate of flow, applied stress and deformation, its rheological behaviour during extrusion for tread and on calendering for gum strips, and to obtain the correct blend of polymers to achieve the desired product characteristics.

The dispersion of carbon blacks, powdery chemicals, other inorganic fillers and so on, in the rubber matrix may require dispersing agents during the preparation of a compound to achieve good processing properties, lower energy costs (for mixing) and the ultimate desired effect for the finished materials. These may include wood rosin, synthetic resins, process oils, pine tars for high-carbon-black-filled compounds, and particular plasticisers for synthetic elastomers. Different kinds of process aid (PA) are used in rubber compounding to improve processability (e.g., fatty-acids, esters of fatty-acids, fatty-acid metal soaps, and low-MW resins).

At the initial stage of mixing a compound, high shear stress and high energy are necessary, but this can be significantly reduced using chemical peptisers, which reduce the viscosity of the basic NR *via* mechanical breakdown. This results in a reduction in energy consumption during mixing and improves the uniformity of the compound by increasing the efficiency of mixing. Some synthetic elastomers can also be treated with peptisers to reduce viscosity, but many can only be heat-treated during mixing before the incorporation of ingredients. Good processing and mixes with good physical properties are achievable when all the compounding ingredients are well dispersed, and a homogeneous compound is prepared for flawless processing.

Adhesion between the used/old tyres and the new tread is the most essential requirement of an elastomer compound, which must be strong enough and should provide the desired strength and durability during service [2].

1.2 Elemental criteria for the selection of rubbers for tyre retreading

The criteria for the selection of elastomers and additives to be processed for retreading are decisive in the choice of the right kind of rubber and the principal ingredients to meet the desired physical parameters. The essential properties considered are as follows:

- Processing distinctiveness
- Moulding and curing characteristics
- Service condition and durability
- Service temperature range
- Duration of service
- Bondability with old tyre
- Physical properties such as mechanical strength, abrasion, cut and cut growth resistance
- Resistance to water
- Oil/chemical resistance (for specific uses)
- Resistance to environmental degradation
- Compound economics

1.3 Rubbers used for tyre retreading

1.3.1 Natural rubber

NR is the most widely used basic elastomer in all categories of the tyre retreading industry. NR has excellent mechanical and processing properties, with excellent green tack and adhesion characteristics with worn tyre. This rubber is largely used in truck tyres because of its road grip and low heat build-up properties.

NR usually contains about 93% of NR (also simply termed hydrocarbon), plus certain inorganic salts and organic materials. The chemical name for NR is polyisoprene, polymer chains of which have the perfect *cis*-1,4 structures that give improved mechanical properties. The high gum tensile strength in this elastomer is due to crystallisation on stretching of its chain. When lower tensile strengths are demonstrated by elastomers that do not properly crystallise on stretching, rubber mixes with an adequate quantity of reinforcing materials are required to increase the tensile strengths. The tensile strengths of different rubbers that may be utilised for tyre retreading are illustrated in Figure 1.1.

NR can contain proteins, sugars, natural rosins and so on. In the grades of NR appropriate for use in tyre tread compounds, the acetone-soluble content must be

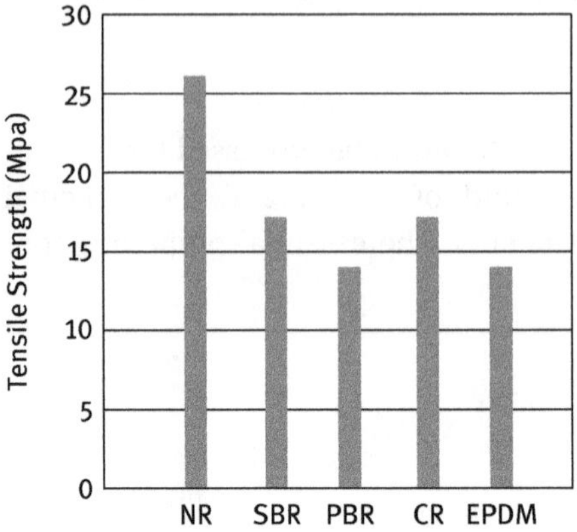

Figure 1.1: Tensile strengths of different rubbers.

in the range of 2.5–4.5%, nitrogen in the range 0.4–0.6%, with an ash content of 0.4%. A precise analysis of NR is provided in Table 1.1.

Table 1.1: Analysis of NR.

Composition	Composition (%)
Rubber hydrocarbon	93.3
Protein	2.8
Acetone extracted matter	2.9
Ash	0.4
Moisture content	0.6

1.3.1.1 Properties of natural rubber

NR is composed of polyisoprene chains with a *cis*-1,4 structure. NR with Mooney viscosities in the range of 60–75 Mooney units is suitable for use in tyre retreads.

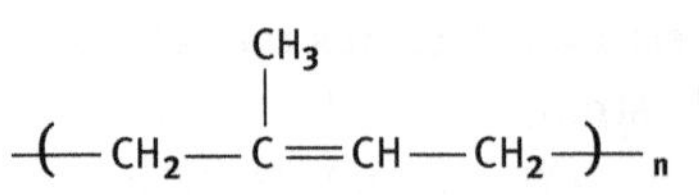

Figure 1.2: Structural formula of NR repeat unit.

A higher viscosity indicates a higher MW, leading to difficulties in processing during mixing, extrusion and calendering.

Having the correct viscosity of raw rubber is essential to making flawless extruded profiles for tyre tread and for the calendering process of tread gum and cushion gum sheeting. Achieving perfect processing behaviour depends predominantly on the viscosity and breakdown behaviour during mastication and mixing of the composition. To reduce complexity, blending of various rubbers from a stock of NR will solve the problem.

In order to control the variations in the quality of NR, several established approaches abide by the technical specifications framed by the International Organization for Standardization (ISO) in 1964 for the production of technically specified rubbers (TSR) such as standard Malaysian rubber (SMR), standard Indonesian rubber, standard Sri Lankan rubber, standard Thai rubber and Indian standard natural rubber. These standards provide the specific limits for dirt, ash, nitrogen, volatile matter and plasticity retention index (PRI), among others. A comprehensive specification for TSR is provided in Table 1.2.

Table 1.2: Specification for TSR.

Parameters	TSR CV	TSR L	TSR 5	TSR 10	TSR 20	TSR 50
Dirt content (maximum %)	0.05	0.05	0.05	0.10	0.20	0.50
Ash content (maximum %)	0.60	0.60	0.50	0.75	1.00	1.50
Volatile matter (maximum %)	0.80	0.80	0.80	0.80	0.80	0.80
Nitrogen (maximum %)	0.60	0.60	0.60	0.60	0.60	0.60
PRI (minimum)	60	60	60	50	40	30
Initial Wallace plasticity (minimum)	–	30	30	30	30	30
Mooney viscosity (ML, 1 + 4, 100 °C)	60 ± 5	N/A	N/A	N/A	N/A	N/A

Constant-viscosity (CV) NR (between 45 and 55) or in the range of 55–65 Mooney units (e.g., SMR 5 CV), can be produced by treatment with hydroxylamine salts at the wet stage. This grade can be used in cushion gum, tread gum or in rubber cement compounds that give CV grades within the stable viscosity level of 45–55 or in the range of 55–65 Mooney units.

The PRI of raw rubber is a test used to assess the level of resistance to the oxidation of NR. The rapid plasticity number of a sample of raw rubber can be used to describe the ratio of the plasticity after heating for 30 min at 140 °C in air to the original plasticity. This is expressed as a percentage of the rapid plasticity number before heating in a Wallace Rapid Plastimeter. Higher PRI values indicate a high resistance to breakdown, and indicate the presence of metallic contamination in the rubber, that can accelerate oxidative degradation. Even trace amounts of metals such as iron, copper and manganese may act as catalysts and elevate the oxidation of NR. The higher PRI value also implies mastication at higher temperature, and can affect the ageing of the rubber compound during service.

Improved grades of NR, such as pale crepes, RSS-1 (ribbed smoked sheet) and SMR 5 have a PRI value greater than 70 [2].

The ISO standards specify six different grades of NR according to a set of rules that feature in the national standards of individual countries technical specifications (TSR). They are graded on technical properties, rather than on visual distinctions. For each grade, instruments test the dirt content, ash content, volatile matter content, nitrogen content, plasticity and colour. A conformity of testing follows from the test methods prescribed by ISO [3].

TSR CV, as well as those for stabilised high-quality latex rubber and TSR 5, 10, 20 and 50, are illustrated in the Table 1.2.

1.3.1.2 Derivatives of natural rubber

1.3.1.2.1 Superior processing natural rubber

Superior processing (SP) NR is produced by coagulating a mix of 20% vulcanised latex and 80% unvulcanised rubber. Blending is performed at the latex stage and it is dried in the usual way. These modified grades of NR are utilised as PA for rubber compounds. They help in the retention of shape and in achieving uniformity in the gauge thickness of the extruded and calendered materials by minimising swelling and shrinkage occurring immediately after these steps.

The commercial grades available are SP rubber PA 80 (a mixture of 80% vulcanised and 20% unvulcanised latex) and PA 57 (57% vulcanised rubber and 30 parts process oil). These SP rubbers are useful for blending with natural or synthetic rubber stocks at a level of 5–10 phr (parts per hundred of rubber) in order to achieve such SP properties as (i) faster extrusion and calendering rate, (ii) better dimensional stability in the extrusion of tyre tread, (iii) less swelling, and (iv) reduced porosity of the extruded and calendered profiles. It is particularly useful when high percentages of elastomers are used in a mix.

1.3.1.2.2 Epoxidised natural rubber

Epoxidised natural rubber (ENR) is produced by chemical modification of NR latex by reaction in hydrogen peroxide and formic acid at 60–70 °C. The latex is neutralised with the application of heat before coagulation, then washed and dried in hot air. This chemically modified new class of elastomers provides the high level of reinforcement achieved from using silica to mixes without the addition of a silane coupling agent. Its significant characteristics are improved resistance to oil and a reduction in gas permeability when used in tyre inner tube compounding.

ENR can be used as a compatibiliser when blends of rubbers are used in compounding. It also acts as a compatibiliser when a high proportion of small-particle-size carbon black or fine-particle silica is incorporated into a rubber mix. Commercially available ENR include ENR-25 and ENR-50 mol%. In order to obtain a clear processing advantage, a dosage of 10 phr ENR-50 mol% may be useful in tyre tread compounds blended with NR/SBR. A lower concentration of about 2 phr can assist in improving adhesion between rubber tread compound and plated steel cord (used to increase the durability of the tyre).

In tyre tread compounding, properties such as wet grip characteristics and rolling resistance may be improved by incorporating ENR in the correct proportions. When blended with SBR, epoxidised NR shows improved bondability, processability, resilience and stiffness. With increasing ENR content, the scorch time and curing time of the compound may favourably decrease [2].

1.3.1.2.3 Chlorinated rubber

Chlorinated rubber is chemically modified NR, and in small proportions finds use in tread gum, cushion gum and cement formulation to promote tack, which will help in fabrication during the retreading of tyres. The chlorination process is performed in solution, or in latex with gaseous chlorine. The chlorine content of the product is usually about 65%. It is available commercially in powder form, and is thermoplastic in nature with a pale cream colour. This material can be used in rubber compounding to enhance the adhesion of stocks [2].

1.3.1.2.4 Ingredients for natural rubber compounding

Dried NR sheets are believed to be adequately protected against oxidation, but this protection may be degraded by excessive washing and heating during processing. Due to the unsaturated character of the NR chains (with a double bond in the main chain), NR is somewhat susceptible to the attacks of atmospheric oxygen, ozone and ultraviolet (UV) light. Through the addition of suitable antioxidants into the rubber mix during compounding, protection can be improved.

Being non-resistant to oils, NR swells when exposed to aromatic, aliphatic and halogenated hydrocarbon solvents. Raw NR is a natural substance, and may degrade at different stages of processing. Its gradation as a raw rubber is classified by factors such as the; presence of wooden bark and sand particles, and active metal contamination (which may initiate oxidation), and care must be taken during the washing of coagulants during processing as this may affect the properties of the dry sheets (naturally occurring antioxidants, if present, may be removed by over washing).

1.3.1.2.5 Chemical peptisers

Different grades of NR in bale form are commercially available with high Mooney viscosity levels. It is essential to masticate the rubber to a particular CV before successfully processing. Viscosity reduction is followed at the processing level by mechanical mastication in a two-roll mill or in an internal mixer. To speed up the initial mastication process, it is appropriate to add a small dosage of chemical peptisers such as the zinc salt of pentachlorothiophenol (Renacit – VII, Bayer; 0.2–0.5 phr). Peptisation activation will be increased by raising the mill temperature in the range of 50–80 °C.

1.3.1.2.6 Metal oxides, silicates and so on

Mineral fillers can yield better dimensional stability and improve extrusion and calendering properties. In the extrusion of tyre tread profiles and calendered cushion or tread gum compounding, the use of precipitated silicas, precipited calcium carbonate and other mineral fillers such as calcium and aluminium silicates and clays can lead to improved processing. These fillers are particularly useful for gum profiles/cement compositions. In combination with reinforcing carbon black, a certain proportion of precipitated silica can improve adhesion properties, as well as increasing the abrasion and tear resistance of the tread compound.

1.3.1.2.7 Zinc oxide

The addition of zinc oxide in sulfur cure compositions provides effectiveness of organic accelerators, so the inclusion of metallic oxide and an organic activator is necessary. Zinc oxide (4–5 phr) is added into the rubber compound in combination with a small proportion of stearic acid (1–1.5 phr). This is effective in formulations based on NR, PBR, SBR, EPDM and so on. In sulfur-curing rubbers, zinc oxide increases the efficiency of vulcanisation, reduces the time for vulcanisation and improves the properties of the vulcanisates. It is also used as a crosslinking agent for CR.

Zinc oxide active, a distinctive grade of zinc oxide, is effective at smaller dosages (2–2.5 phr), because of its smaller particle size. It also increases the degree of crosslinking, but slows the initiation of cure, although with no influence on the effective time of cure [4].

1.3.1.2.8 Process aids

Petroleum process oils at a level of 5–10 phr can help in the process of incorporating fillers during mixing in mills. They act as plasticisers while processing, reducing the viscosity of the polymer. Aromatic type oils give excellent processing ability for NR-based compounds.

PA, as factices, contributes greatly to the processing behaviour of NR- and synthetic rubber-based compounds. Brown factices improve extrusion and calendering quality, as well as dimensional stability in the extrusion of treads, when used in dosages within 5–10 phr. They also facilitate the dispersion of powdered fillers by controlling the ‘nerve’ of the elastomers.

There is a tendency to ‘stick to calender rolls’ when processing cushion gum and tread gum mixes, in particular for NR-based low-hardness compounds. In order to deal with this problem of sticking to rolls, fatty-acids, and derivatives such as the zinc salts of high-MW unsaturated fatty-acids (Aktiplast, Rhein – Chemie) or low-MW polyethylenes, can be used as additives [2].

1.3.1.2.9 Antioxidants/Antidegradants

Antioxidants/antidegradants are selected for tread mixes by considering the severity and type of service exposure conditions, such as heat, flexing, oxygen and ozone. In general, in the antioxidant group, polymerised dihydroquinoline [trimethyl quinoline (TMQ)], is widely used in combination with antiozonants in tyre tread mixes.

Microcrystalline (MC) waxes (Antilux) obtained from higher-MW petroleum residuals are used in tyre retreading compounds to protect against ozone where high flexing is required. Unlike paraffin wax, this wax does not crack when bloomed to the surface during flexing, and provides a protective antiozonant coating. It also performs better at elevated temperatures. MC waxes have a predominantly branched structure of irregular smaller crystals that melt from 57 to 100 °C. Its mobility intensifies at higher temperatures, so its rate of migration to the surface increases. A combination of wax and a small amount of chemical antiozonant results in a synergistic effect in ozone resistance during the service of the tyre tread.

Antiozonants such as *p*-phenylenediamines (DPPD) and alkyl-aryl-*p*-phenylenediamines (PPDA) offer excellent protection in ozone-sensitive applications, and work as an antiflex cracking agent under cyclic deformation when the tread is in use. In combination, an antioxidant such as TMQ or phenolic antioxidant with PPDA give excellent protection against ozone cracking and mechanical fatigue in tread. Proper blending of acetone-diphenylamine condensate (Nonox BL) with DPPD or PPDA improves antiflex cracking and heat ageing properties [5].

1.3.1.2.10 Tack aids, tackifiers

Stocks used for tread gum, cushion gum and cement applications in the tyre retreading process need to be soft and tacky. Tread gum and cushion gum sheeting of required thickness is prepared by a calendering process. The incorporation of resins such as coumarone–indene, petroleum or wood rosins will help to achieve the desired tack.

In addition, these natural and synthetic resins are helpful in the dispersion of dry powdered fillers.

In order to promote adequate tack to facilitate building of the worn tyre for retreading, the incorporation of pine tar (produced by the distillation of pine wood) in the rubber composition (3–5 phr) provides satisfactory results. This tackifier also helps in dispersing high carbon black filled mixes, and can be used in higher proportions in gum/cement stocks to improve adhesiveness.

The incorporation of resorcinol resin and hexamethylenetetramine in addition of silica to tread and cement formulations will further intensify bonding strength.

The system increases adhesion by improving wetting and hydrogen bonding between the rubber compounds and the substrate. Silica plays a role in preventing the loss of adhesion as the tread ages or at elevated temperatures.

Before the addition of an accelerator during the mixing cycle, resorcinol and precipitated silica (4–6 phr), in combination with hexamethylenetetramine coated with silica (1.5–2 phr), can be added directly to the rubber mix. To obtain the best adhesion, the curing system for the rubber matrix is formulated with the delayed action of an accelerator such as sulfenamide and sulfur [6].

1.3.1.2.11 Solvents and chemical bonding agents

In the preparation of rubber adhesive/cement intended for the building up operation, it is necessary to swell the mixes in hydrocarbon solvents such as toluene, solvent naphtha or trichloroethylene. A solution of desired viscosity is prepared, and the prepared cement is then applied on the worn and prepared tyre surface.

1.3.1.2.12 Accelerators

NR-based tyre retread mixes or blends with PBR/SBR used in soluble or insoluble sulfur vulcanisation systems use an organic accelerator together with a metallic activator (zinc oxide and stearic acids) to give the desired results. In general, for tyre tread mixes where NR/SBR/PBR blends are used, sulfanilamide (cyclohexyl benzothiazole sulphenamide, CBS) accelerator alone can be used. This type of accelerator is safe at processing temperatures, and is slightly faster at the curing temperature than thiazoles (dibenzthiazyl disulphide). When a high proportion of reinforcing furnace black is incorporated into the tread composition to improve the property of reinforcement, a delayed-action accelerator (delayed compared to CBS) such as butylbenzthiazyl sulphenamide or benzthiazylsulphenmorpholide (BSM) will impart excellent processing safety and outstanding curing characteristics. The accelerator BSM is also effective with high-silica filled compounds.

Due to the high ‘heat history’ of the compound and the processing at higher temperatures during the mixing and extrusion of tread, ‘scorching’ or premature vulcanisation might occur. This phenomenon can also prevail in stocks containing high proportions of low-structure carbon black in NR (also with other highly unsaturated rubbers). To control this problem during processing, a small dosage of retarders/ prevulcanisation inhibitors such as nitrosodiphenylamine, salicylic or benzoic acids (0.25–0.50 phr) can be added to produce a ‘scorch safe’ compound. However, the addition of these chemicals may increase the curing time.

It is worthwhile to use the same category of accelerator/s in all categories of mixes for the preparation of components for retreading assembly; a slightly higher dosage may be used in cushion gum and cement compositions because of the slower heat penetration in those areas during vulcanisation [7].

1.4 Styrene-butadiene rubber

In tyre tread compounding, the general-purpose synthetic copolymer of styrene and butadiene is effective for retreading all categories of tyre when blended with NR. A random copolymer from the aforesaid monomers is commercially available as one of two major types: (i) emulsion SBR and (ii) solution SBR, which are produced by two different methods of polymerisation – emulsion and solution. The two procedures give distinctive final properties. When compared with the emulsion process, the solution method offers a narrower MW distribution with higher content of *cis*-1,4 structure and lower glass transition temperature (T_g) to the elastomer. However, the solution SBR offers higher flexibility than emulsion grades (a significant property for tyre applications).

Generally, the monomer ratio for styrene and butadiene when producing SBR is 30:70. Polymerisation is carried out in 'cold' or 'hot' processes. The cold grade polymerisation is conducted at 4 °C, and the hot grade at 42 °C.
Tyre retread rubber mixes are chosen for the following reasons:
- Good processability
- Excellent green tack and good green strength
- Low rate of cut growth
- Good resistance to skid
- Low hysteresis
- High wear resistance
- Good rolling efficiency
- Stability to thermo-oxidation
- Economical compounding

The important factors are the dynamic properties that are relevant to different service requirements, including mechanical characteristics such as tensile strength, elongation at break and resistance to abrasion.

Tread rubber compounds based on solution-grade SBR show better dynamic properties, including higher elasticity, high flexibility and lower heat build-up. These important properties can be requirements in passenger car, bus tyre, and truck tyre retread compositions. Solution SBR in tread can also impart better grip, lower rolling resistance and longer service life.

SBR produced with these two methods exhibit certain advantages and disadvantages in their properties. These characteristics are modified by making blends with NR to achieve a balance between the properties. Blending of a solution grade of SBR with emulsion grade can improve its processability due to a broadening of molecular macrostructures and degree of branching (MWD).

Most commonly, 23–25% of bound styrene is found in SBR produced by the emulsion process. With increasing styrene content, processability improves, and also the tensile strength, but the resistance to low-temperature and wear deteriorates. Different levels of MWD assist with processing behaviour and product properties [8].

1.4.1 Properties of styrene-butadiene rubber

SBR molecules are composed of styrene-*co*-butadiene chains, with the molecular structure shown in Figure 1.3.

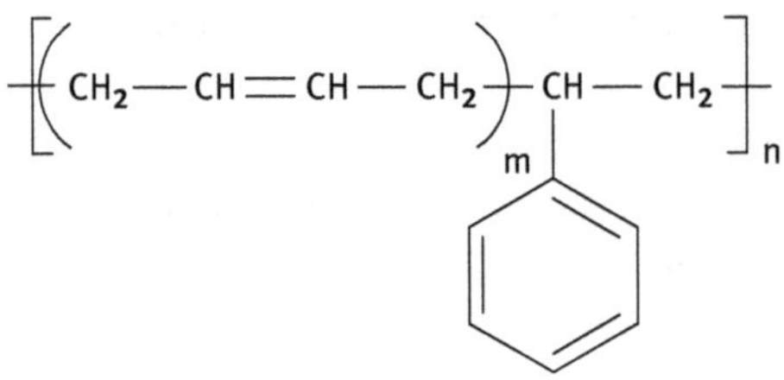

Figure 1.3: Molecular structure of SBR.

The double bond is present in the butadiene molecule and can be present in the following forms:
- *Cis*-1,4
- *Trans*-1,4
- Vinyl-1,2

The International Institute of Synthetic Rubber Producers' criterion on the classification of SBR is as follows:
- 1000 series – hot polymerised, non-pigmented
- 1500 series – cold polymerised, non-pigmented

Solution SBR is most suitable for designing tyre compounds for tread and sidewall. It is appropriate for its inherent hysteresis and abrasion resistance effects, which improve the performance of the tyre by controlling the elastomer's T_g through adjustment of the styrene and vinyl content.

With various bound styrene contents and different proportions of 1,2-addition of solution SBR with a view to increasing the content of phenyl and vinyl substituents on the basic hydrocarbon chain, the processability of the rubber compound is improved. Its rebound resilience decreases, the grip property of the tyre tread improves, and road holding, including skid resistance on wet roads, is better; however, wear resistance declines. Many grades of SBR are cost-effective and available commercially [9]. The basic specification of SBR 1502 is provided in Table 1.3 [2].

Table 1.3: Basic specification of SBR grade 1502*.

Property	Maximum %
Butadiene	75.00
Styrene	25.00
Free organic acid	7.00
Soap	0.50
Specific gravity	0.92
Non-staining antioxidant	1.75
Volatile matter	0.75
Iron content	0.005
Copper content	0.0002
Ash	1.50
Mooney viscosity	58

*SBR 1502 is a cold grade with appropriate reinforcement. This grade can offer very good abrasion resistance and resistance to flex

1.4.2 Styrene-butadiene rubber 1712 grade

SBR 1712 grade is produced by co-polymerising butadiene with styrene in emulsion at 5 °C, using a mixture of rosin and synthetic acid soaps as an emulsifier. The rubber is precipitated from latex using salt and acid techniques.

SBR 1712 with 23.5% bound styrene content is a cold polymerised elastomer, extended with 37.5 parts of aromatic oil, available in 45–55 Mooney viscosity. The aromatic oil used is an effective plasticiser for high-MW SBR. Careful selection and correct blending proportions of this grade with NR provides a tread composition with very good mechanical properties and processing characteristics in an economical way. The characteristic features of this elastomer are its dark brown colour, high abrasion resistance, excellent processability, good scorch properties and trouble-free vulcanisation when blended with NR. A specification for SBR 1712 is provided in Table 1.4.

The SBR 1502 and 1712 grades can be blended easily with natural and other synthetic rubbers. These grades do not require premastication during mixing. They give good dimensional stability and can offer good processability during extrusion of tread and calendering of gum sheeting, but they have rather poor physical properties.

Their improvements in mechanical properties such as higher tensile strengths and higher abrasion resistances can be achieved when properly formulated with reinforcing fillers such as reinforcing carbon blacks and with the addition of precipitated silicas in the mixes. Hot polymerised SBR grades are prone to difficulties in processing in general, although they offer low shrinkages and good calendering behaviour, but again they show lower physical properties [10].

Table 1.4: Specification of SBR 1712.

Property	Maximum %
Bound styrene	23.50
Specific gravity	0.95
Free organic acid	5.0
Antioxidant	1.30
Volatile matter	0.50
Ash	1.0
Iron	0.006
Copper	0.0002
Oil	37.5
Mooney viscosity	45–55
Colour	Dark brown

The mechanical properties of rubber tread compounds improve when polymers are used for their production that have higher MW. The strength properties and wear resistance improve and the heat build-up on cyclic stressing decreases. When the MW of the polymer is increased, its viscosity also increases, so the processability gradually deteriorates to an unacceptable level.

The physical properties of the rubber vulcanisates produced from oil-extended rubbers are somewhat poorer than the properties of those made from unextended rubbers, but they are cheaper in terms of compounding.

SBR-based compounds are similar to those of other unsaturated hydrocarbon rubbers. Furnace blacks such as intermediate super abrasion furnace (ISAF) carbon black or high abrasion furnace carbon black grades are usually added to tread compounds to enhance the physical properties. Building tack is lower, which is a disadvantage of using SBR on its own in compounds.

In comparison with NR, emulsion SBR, when used in tread compounds, has the following properties:

- Better wear resistance under severe conditions
- Better ageing resistance
- Greater resistance to crack initiation and growth and fatigue with repeated small strains

When compared with NR, the lower resistance to crack growth and fatigue at larger deformations are among the disadvantages of emulsion rubbers, as also are lower tear strength, and poor dynamic properties such as lower rebound resilience, higher hysteresis and higher heat build-up on cyclic deformation. Excessive heat build-up under dynamic conditions makes the use of SBR for large truck and dumper tyres unfeasible.

The processing characteristics of SBR-based compounds are similar to those for NR. Because of its inadequate green tack, it is not generally used on its own for retreading as a base polymer. To improve the green tack of SBR with the substrate, wood rosin/ coumarone – idene resins can be incorporated into mixes to help increase the tack and physical properties of the compound. Blending with NR (NR/ SBR ratios of 80:20 or 60:40) in practice improves the processing characteristics of the stocks, and leads to a proper balance of properties.

SBR stocks have less tendency to scorch during processing because of their low unsaturation in the polymer chain and slower curing rate than NR. Its low unsaturation gives better heat ageing and better heat-resistance qualities. The elastomers also have better abrasion properties and better crack initiation resistance than NR. These rubbers also have low water absorption, which is an advantage when the tyres operate generally in wet areas [2].

1.4.3 High-styrene resins

High-styrene resins (HSR) are copolymers produced by an emulsion polymerisation method that applies the same technique as used in producing SBR, by reducing the butadiene content to 15% and the styrene content to 85%. It is coagulated on polymerisation and filtered, and is commercially available in the form of thermoplastic, free-flowing white granules.

HSR may be used with a lower proportion (5–10 parts per hundred) of basic elastomer to facilitate processing and reinforce NR and synthetic rubber to some extent to increase hardness, modulus, abrasion, tear resistance and flexibility. This polymer is compatible with NR, SBR, PBR, CR and EPDM rubbers; to achieve optimum results, the resin must be properly blended with the basic polymer at the initial stage of mixing. It starts to become softer at 50 °C because of its thermoplastic nature, but to achieve better results in blending with the basic elastomer, the mixing temperature should be about 80 °C.

Adding this polymer in a small proportion to the base rubber compound used for retread formulations can substantially improve the processing behaviour during mixing, extrusion and calendering. It also improves the surface appearance of intricate tread designs, particularly with the precured tread used for the cold curing system [6].

1.5 Polybutadiene rubber

PBR are manufactured using the solution polymerisation method with complex organometallic catalysts. There are two main categories of solution-polymerised polybutadienes that have become significant commercial rubbers (Figure 1.4):

- High *cis*-polybutadiene (1,4-polybutadienes)
- Medium *cis*-1,4 polybutadiene

$$\left(CH_2 - CH = CH - CH_2 \right)_n$$

Figure 1.4: General formula of PBR repeat unit.

In tyre retreading applications, PBR is not used alone as the base polymer because of its poor gum strength and very low building tack. Thus adhesion to substrates is similar to other diene rubbers, and the addition of resins and a compatible adhesion promoter to mixes can promote tackiness and facilitate the fabrication process during retreading. The preferred compounding ingredients – PA, curatives and so on – are similar to those for other general-purpose rubbers.

The proper blending of PBR with NR in appropriate proportions improves tyre tread properties such as adhesion, abrasion and chipping resistance. When also blended with SBR together with NR, it can provide improved crack growth resistance [2].

1.5.1 Effective properties of polybutadiene rubber for retread applications

PBR exhibit a lower tensile strength than SBR, but this difference decreases at elevated temperatures. They retain their strength well at elevated temperatures.

The outstanding properties of PBR, including very high resilience, excellent low-temperature flexibility and excellent abrasion resistance, are significant when compared to NR, SBR or EPDM polymers. When compared with NR the rebound resilience of PBR is better at low temperatures, but the NR has a higher rebound resilience than PBR at higher temperatures. Compared to NR, the dynamic modulus and loss modulus are higher in butadiene rubber. The lower tensile strength and tear strength of polybutadiene prevent its use in tyre applications at higher concentrations. In view of its lower tensile strength, PBR is usually blended with NR to improve this property considerably.

In NR-based formulations where SBR are incorporated into the tread composition for retread, observations reveal the formation of small cracks in the grooves of the tread in truck, bus or even passenger car tyres. These cracks may lead to a breaking or separation effect during the service life of the tyre. The PBR elastomer's excellent resistance to tread groove cracking is an important component of tread compounds for all categories of tyre.

Inclusion of this polymer in the design of tread formulations not only improves the crack resistance, but also improves resistance to wear. It may also impart uniform radial wear of the tyre, particularly in heavy-duty tyres. As a result it can increase the service life of a tyre on an average basis.

Because of its excellent low-temperature properties, *cis*-polybutadiene can maintain its resilience at −40 °C. To achieve resistance to stiffness at low temperatures, this polymer is therefore blended with NR and SBR for retreading applications.

A combination of NR and SBR with polybutadiene is used to explore the specific properties of this elastomer in retread, such as high resilience, good crack resistance, high abrasion resistance, good ageing and weathering resistance, and overall good low-temperature flexibility when the tyre is used in low-temperature areas. PBR can also provide good dimensional stability, and can improve the flow properties that are particularly important for high-rubber-content compounds processed in an extruder or in a calender machine.

The NR/PBR blend with a ratio of 80:20 improves processability by contributing to increasing extrusion speeds in low-temperature operation, and can give a smoother, softer, flexible, low-porosity tread profile with minimum thickness variations.

PBR grade 1220 has a remarkable property of high wear resistance. This grade of PBR is easily compatible with NR and SBR in any proportions. When used in mixtures with other rubbers, this grade retains their individual properties and adds high elasticity, wear and cold resistance.

These rubber products are subject to dynamic applications, so resistance to crack initiation and growth, heat stability, low heat build-up and high wear resistance are essential requirements. PBR/NR blends in the correct proportions will provide long operational service life in tyre retread applications.

The property of low water absorption is a necessary feature for tread mixes where the tyre will be used in constantly wet areas [10]. If the retread tyre is to be used in snowbound areas, this rubber is primarily used in tyre tread compositions because of its high elasticity, exceptional wear resistance and high cold resistance. A typical specification for PBR grade 1220 is provided in Table 1.5 [9].

Table 1.5: A typical specification for PBR 1220.

Properties	
Cis-1,4	97%
Specific gravity	0.93
Ash content	0.5%
Organic acid	0.1%
Mooney viscosity (ML, 1 + 4) 100 °C	45 unit

For comparison, the physical properties of different elastomers using ISAF grade reinforcing carbon black at the 50 phr level, with added PA and curatives, are shown in Table 1.6.

Table 1.6: Comparison of properties of tyre tread composition using NR, PBR and SBR.

Property	NR	SBR	PBR
Hardness (Shore A)	68	65	64
Tensile strength (MPa)	24	20	16
300% modulus (MPa)	12	9	8
Elongation at break (%)	550	480	500
Rebound resilience (%)	70	60	75
Heat build-up (°C)	15	22	15

1.6 Polychloroprene rubber

Neoprene is the common generic name for the chloroprene polymer produced by the emulsion polymerisation method from the monomer 2-chloro-1,3-butadiene in the presence of emulsifying agents in water, catalysts, modifiers and protective agents. Polymerisation is carried out under controlled temperature and time. During the polymerisation process, 1,4-addition (at about 98%) and some 1,2-addition (approximately 1.5%) take place. Although the 1,2-addition is small, it must be moderated with care, as it typically has a marked effect on the final properties of the polymer. Some chlorine is also available on the polymer chain (in the allylic form) during the course of 1,2-additions because it has been used as curing location on the polymer. The emulsion is subjected to coagulation and then dried to obtain the polymer (Figure 1.5).

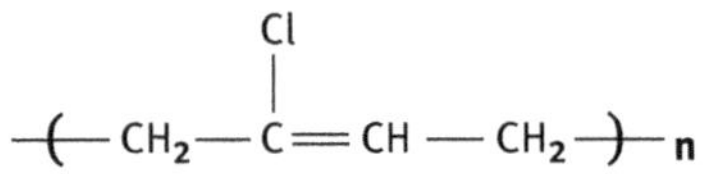

Figure 1.5: Structural formula of polychloroprene rubber repeat unit.

Polychloroprene rubber is available for industrial use in the following two dominant varieties:

- General purpose grades
- Adhesive grades

1.6.1 Properties of polychloroprene rubber

CR have high mechanical strength, excellent bonding strength, a well-balanced combination of weatherability and resistance to ozone, and heat, flame and oil resistance. CR are generally classified according to Mooney viscosity and crystallisation rate. Mooney viscosity is essentially determined by MW, with a higher Mooney viscosity generally corresponding to a larger MW. When vulcanised, grades achieve a higher Mooney viscosity; they increase in strength, but their viscosity increases. Their propensity to scorching during processing leads to poor processability in extrusion while making tread rubber for retreading purposes.

This elastomer is usually classified into three types:
- Mercaptan-modified (excellent extrusion and calender properties).
- Sulfur-modified (its MW can be reduced through mastication, which breaks down the molecules, as with NR, to obtain a low-viscosity compound; this rubber, when vulcanised, exhibits high elasticity).
- Xanthate-modified vulcanised products (high mechanical strength; inexpensive blends based on a high loading formulation show high antivibration performance).

Mercaptan-modified CR gives better heat resistance than sulfur-modified or xanthate-modified CR, and sulfur-modified CR is available with less heat build-up.

In the search for oil and high weather resistance in tread compounds for tyre retreading, blending of CR with NR in the correct proportions can offer the desired properties. The adhesive grades of CR may be selected to provide adhesives for tyre repair for room temperature curing applications.

There are many properties of CR that can be exploited in blending with other rubbers. To improve elasticity and low-temperature flexibility, CR can be blended with NR. Blending with PBR may improve the brittleness point of rubber compounds at low temperatures.

The gum tensile strength and tear resistance of CR is high. Adding reinforcing carbon blacks can achieve a further increase in these properties. To obtain a higher tear strength, reinforcing silica may be incorporated into the compounds for tyre tread uses.

When produced correctly, compounded CR has the characteristics of excellent ageing, weathering and ozone resistance. It is moderately resistant to oils and a number of chemicals. These rubbers have good mechanical properties and abrasion resistance, very good hot air resistance, and very good resilience. Its tendency to crystallise after curing shows a high level of resistance to flex cracking of treads.

The addition of petroleum oils, resins and plasticisers into CR based compounding has advantages. Plasticisers include dibutyl phthalate, dibutyl thioglycolate (Plastikator 88) and ether thioether (Plastikator OT). These need to be incorporated into polychloroprene-based mixes to obtain the following benefits when designing tread compounding: (i) dispersion of fillers, (ii) facilitation of processing, and (iii) increased adhesion.

Metallic oxides such as magnesium oxides and zinc oxides are used in CR compounding to assist in processing safety, rate of cure and quality of the cured tread. Accelerators such as sulfanamides, thiazoles, thiurams, thioureas and sulfur may be added in lower dosages as curing agents in retread compositions. Fast curing accelerators such as ethylene thiourea (NA 22) and diorthotolylguanidine can be used in CR retread recipes.

Polymers with the intrinsic property of flame retardation are suitable for the manufacture of products that need to be flame proof, as when a tyre is used in a fire hazard environment. This property can be further improved by adding chlorinated paraffins, antimony oxides and zinc borate into mixes in the correct proportion [11].

1.7 Ethylene propylene rubbers

EPDM is a terpolymer polymerised from ethylene, propylene and a small percentage of diene (Figure 1.6). The diene provides unsaturation in the side chain or pendant in order to offer sites for crosslinking.

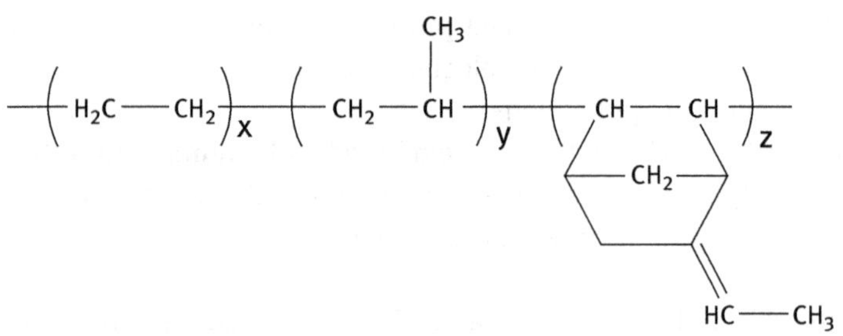

Figure 1.6: Structural formula of EPDM polymer.

Commercially available EPDM usually contains 4–5% by weight of diene and 30–70% is *ethylene*. The Mooney viscosity of EPDM is in the range of 45–120 Mooney units. It can be blended with NR in some proportion to achieve some of the specific properties needed for use in tyre tread compounds.

A high proportion of ethylene or propylene in rubber offers higher tensile strength and elongation. On increasing the ethylene content, the end product shows increased hardness, high green strength and higher tensile strength, as well as enhanced rebound resilience. It offers outstanding resistance to oxygen, ozone and UV rays exposure because of its saturated main chain.

The following are some of the special properties of EPDM:
- Very low specific gravity
- Very good mechanical properties, very high extendibility
- Impermeability to gases and water
- Very low moisture absorption
- Excellent heat and ageing resistance
- Resistance to UV and oxidative cracking due to ozone

- Excellent low-temperature flexibility (very low brittle point and T_g)
- Excellent resistance to chemicals
- Good resistance to animal and vegetable oils

To enhance the physical properties of EPDM rubbers, the incorporation of reinforcing carbon blacks/reinforcing white fillers in compounding is essential. A graphical representation of the reinforced tensile strength of EPDM rubber is presented in Figure 1.1.

EPDM rubber is not suitable on its own for use in tyre tread compounding because of its lack of tackiness and not very high physical properties. Insufficient tack of the stocks will impede fabrication with old tyres. Blending with NR and the use of tackifying resins are therefore required in order to enhance the physical properties and the tack of the green compound for retread purposes. Common rubber accelerators are suitable for curing unsaturated terpolymer EPDM, but at a higher dosage than generally used for unsaturated rubbers. The properties of this elastomer are detailed in Table 1.7 [12].

Table 1.7: EPDM specifications.

Mooney viscosity ML (1 + 4) at 125 °C	Propylene content (% by weight)	Ethylene content (% by weight)	Ethylidene norbornene (% by weight)	Volatile matter max (% by weight)	Ash content max (% by weight)
75–85	38–42	52–56	4.5–5.5	0.70	0.20
70–80	26–30	64–68	4.5–5.5	0.70	0.20
44–54	33–37	54–58	7–9	0.70	0.20
45–60	33–37	59–63	2–3	0.70	0.20
40–50	35–40	56–61	2–3	0.70	0.20

1.8 Reclaim rubber

Reclaiming is the predominant technique for the partial recycling and retrieval of waste rubbers. The standard method of reclaiming is thermal-oxidative or reprocessing of the vulcanised rubber.

Regenerated or reclaim rubber is produced by treating vulcanised scrap rubber such as that from tyres, tubes and various other waste from rubber-made products. Rubber scraps are ground, then, through the application of heat, devulcanisation of the polymer is achieved. This method is effectively depolymerisation, but without the removal of combined sulfur. The outcome of the material – reclaim – can be treated like an elastomer and used with virgin rubber in compounding then vulcanised to form custom rubbers.

While reclaiming the vulcanised rubber scraps the MW decreases and chain scission occurs. During depolymerisation of the crosslinked rubber, a break in the molecule occurs.

The following processing steps are carried out to produce reclaim rubber:
- Rubber scraps categorisation and grinding
- Removal of textile fibres and metals from the rubber
- Devulcanisation
- Treatment of the devulcanised rubber by mechanical means

Reclaim rubber can be produced much more cheaply than original rubber. It is therefore used widely as a rubber raw material to facilitate processing and also for economic considerations. It can replace a proportion of new rubber (either natural or synthetic) without having much effect on the physical properties of the retread compound. However, physical properties of the tyre retread compound such as tensile strength, tear resistance, resilience and abrasion resistance will depreciate if the proportion of reclaim rubber used in retread mixes is increased too much.

A fine-grain or extrusion variety of this rubber can also be added in small proportions to tread gum and cushion gum formulations without greatly affecting their physical characteristics. Indeed, reclaim rubber can lead to improved extrusion and calendering properties. The basic specification of reclaim rubber is presented in Table 1.8 [13].

Table 1.8: Basic specification of reclaim rubber.

Properties	Proportion (%)
Rubber hydrocarbon content	50
Carbon black content	28
Specific gravity	1.12
Volatile matter	1
Acetone extract	14 (maximum)
Ash content	8 (maximum)
Mooney viscosity ML (1 + 4) @ 100 °C	25–55

There are various methods to produce reclaim rubber commercially, but the following are the main industrial techniques:
- Digester process
- Pan process
- Reclaimator process

In the process of producing reclaim rubber waste, rubber (vulcanised) is heat-treated for long periods at a temperature of around 200–250 °C. The vulcanised rubber is depolymerised at this temperature, and reclaim rubber is generated. This material can be used again in combination with virgin rubber compounds. The process of regeneration can be speeded up by the addition of chemicals or reclaiming agents such as (i) thiophenols and disulfides and (ii) dixylyldisulfide or alkylated phenylsulfide (usually used for synthetic rubber reclaiming).

1.8.1 Digester process

Rubber scraps cut into pieces to a size of 6–8 mm are made free from fabric, and metals, if present, are separated using a magnetic separator. The stock is then transferred into a digester with water, oils and additives. The digester is heated with steam and the temperature maintained at about 198 °C for 4–12 h with continuous stirring. After heating, the contents of the digester are dumped onto a conveyor and conveyed to an extruder for straining and refining. They are then deposited onto a drum, from which the materials are removed in the form of slabs.

1.8.2 Pan process

This method can be used to reclaim NR, SBR, CR, NBR and isobutylene–isoprene rubber. The raw material produced from vulcanised rubber waste is ground into smaller particles, made fabric-free, then mixed with oil and other additives and transferred to a steam-heated vulcaniser. Steam-heating is conducted at 14 kg/cm^2 pressure for several hours, and devulcanisation takes place. The stock is then unloaded and conveyed to a mill for refining and straining following usual methods.

1.8.3 Reclaimator process

Most commercially available reclaim is produced using the reclaimator process. In this method of reclaiming, fibre-free vulcanised rubber scraps are ground using 30 meshes. They are then heated at a temperature of 200–260 °C, and devulcanisation takes place on rapid heating of the stocks. In order to obtain the desired plasticity, the time used is about 3–6 min, and plasticity is attained by incorporating additives such as plasticisers and catalytic agent in the presence of oxygen.

A specially designed screw extruder with a discharge head with regulated clearance operates as the reclaimator machine. Automatic feeding of rubber crumbs (mixed with oil and chemical agents) through the hopper enables continuous reclaiming. The heat generated by mechanical working of the machine initiates depolymerisation, and the progress of the reaction is controlled by adjusting the clearance of the discharge head. The process time, temperature and therefore the required product quality are closely controlled, and the appropriate cycle time is only 30 min. The depolymerised stocks are then mixed with PA and reinforcing agents, subsequently sheeted and formed into slabs for commercial use.

The significant processing and cost advantages of using reclaim rubber in tyre retreading compounding are as follows:

- Lower breakdown time and mixing time; good dispersion in virgin rubber due to low nerve
- Lower power consumption during mixing
- Low-temperature development during mixing and extrusion of tread
- Faster and uniform extrusion; controls dimensional stability
- Increase and retain tack of the compound
- Low swelling and shrinkage of the tread profile, gum sheeting; through extrusion and calendering
- Higher rate of cure
- Reduced cost of compounding

The normal shrinkage of stock with high virgin rubber content varies from 6% to 10%, and incorporating reclaim rubber into the compound helps to reduce this. The reclaim rubber has less nerve, which results in lower internal heat build-up during the processes of mixing, extrusion and calendering [12].

1.8.4 Crumb/Powdered rubber

Rubber crumb is the material produced by granulating scrap tyres into standardised rubber particles. Crumb rubber, or rubber powder when of a finer consistency, is recycled from old tyres. Some can be used in retread formulations without affecting the physical properties of the rubber compound. Whole scrap tyres can be salvaged as crumb rubber and converted into an effective and economical raw material for retreading. In converting the whole tyre, steel, fabric and any other foreign matter are removed carefully to give a consistent granular formation.

Sorted tyres are granulated by a shredder machine. Further processing is carried out in a cracker mill, and the granules are ground in a grinding mill to reduce the particle size further. To segregate the particles by size, they are then sieved in a vibrating sieving machine fitted with an appropriate mesh for the tyre tread application (a 40–80 mesh is suitable).

A considerable proportion of crumb rubber may be incorporated into tread mixes depending on the type and use of the retread tyre. Finer grades can be used in small proportions in tread gum compounding. Using properly ground rubber crumb in the rubber compound leads to good extrusion and calendering. It also assists in removing trapped air from thicker extruded profiles such as tyre tread (common while extruding thicker profiles). Here, again, like reclaim rubber, higher proportions may affect the physical properties of the tread.

Granulation and grinding can be performed under ambient processing conditions, but there is also a cryogenic method in which liquid nitrogen can be used to freeze chips of scrap tyres at about −80 °C to reduce the size of the granulated rubber particles. The reduction of particle size using this method prevents heat degradation of the rubber, which occurs while grinding at normal temperatures [12].

The distinctive properties of the different elastomers used for the manufacture of tyre retreads are given in Table 1.9.

Table 1.9: Distinctive characteristics of different rubbers used for tyre retreading.

Properties	NR	PBR	SBR	EPDM	CR	Reclaim rubber
Adhesion to substrate	E	G	G	F	E	VG
Impermeability to air and gases	G	G	F	F	G	F
Abrasion resistance	E	E	G	G	E	G
Resilience	E	E	G	G	G	G
Environmental resistance	P	P	P	E	G	F
Ozone resistance	P	P	P	E	G	P
Resistance to UV light	P	F	F	E	G	F
Resistance to flame	P	P	P	P	G	P
Low-temperature flexibility	E	E	G	G	F	G

E: Excellent
F: Fair
G: Good
P: Poor
VG: Very good

References

1. B Banerjee and S.K. Pal in *Retreading of Tyres – Methodology and its Significance in Economics and Environment: Book of Abstracts*, India Rubber Expo – Conference and Workshop, Mumbai, India, 2013, p.189.
2. *Advances in Polymer Coated Textiles*, Ed., G. Akovali, Smithers Rapra Technology Ltd, Shawbury, UK, 2012, p.25.
3. ISO 2000:2003, *Rubber, Raw Natural – Guidelines for the Specification of Technically Specified Rubber (TSR)*, International Organization for Standardization, Geneva, Switzerland, 2003.
4. *Science and Technology of Rubber*, 3rd Edition, Eds., J.E. Mark, B. Erman and F.R. Eirich, Elsevier Academic Press, London, UK, 2005.

5. *Elastomers and Rubber Compounding Materials*, Ed., I. Franta, Elsevier Science Publishing Company, New York, NY, USA, 1989.
6. *Rubber Technology and Manufacture*, Ed., C.M. Blow, Newnes–Butterworths, London, UK, 1971.
7. *Manual for the Rubber Industry*, Farbenfabriken Bayer AG, Leverkusen, Germany, 1972.
8. W. Hofmann in *Rubber Technology Handbook*, Hanser Publishers, Munich, Germany, 1989.
9. *Developments in Rubber Technology – 4*, Eds., A. Whelan and K.S. Lee, Elsevier Applied Science, London, UK, 1987.
10. *Rubber Technology*, Ed., J.S. Dick, Hanser Publishers, Munich, Germany, 2001.
11. *The Vanderbilt Rubber Handbook*, R.T. Vanderbilt Company, Norwalk, CT, USA, 1978.
12. *Polymers in Construction*, Ed., G. Akovali, Rapra Technology Limited, Shawbury, UK, 2005, p.78.
13. *Rubber Engineering*, Indian Rubber Institute, Tata McGraw-Hill Publishing Company Limited, New Delhi, India, 1998.

Bireswar Banerjee

2 Processability, characterisation and properties of tyre retread compounding ingredients

2.1 Introduction

The major requirements for an extruded retread profile intended for retreading (with a view to maintaining consistency) are (i) to be free from porosity, (ii) to be free from blisters and (iii) for the profile to be free from surface imperfections. Uniformity in the shape of the extruded tread section enables proper placing on the buffed contour dimensions of the old tyre that is ready for retreading. The compounding of rubber, its homogeneous mixing, shaping of a tread profile by extrusion, calendering of tread gum and cushion gum sheeting are all steps in the process of achieving a flawless retreading operation.

The most important characteristics required in a retread rubber composition are outstanding adhesion with the old tyre crown, very good tear and cut resistance, and excellent abrasion and ageing resistance after curing. The processing characteristics to achieve efficient bonding of the new tread with the old tyre throughout its length of service, as well as appropriate mechanical properties in the final product and cost-effective economical compounding, must be effectively developed within the factory.

To achieve these aims, a suitable formulation of the rubber compounds, the steps of mixing, extrusion and calendering, and the selection of appropriate ingredients such as reinforcing carbon blacks, precipitated silicas, process aids (PA), accelerator and antidegradant are important decisive factors. Appropriate curing parameters (either cold curing or hot curing) should be followed strictly to achieve the desired properties for the product.

Calendered items required for tyre retreading assembly, such as the tread gum profile and cushion gum sheeting mixes, are processed in a calender machine and have to be of a high quality in terms of lack of porosity, blemish free, consistent and uniform in thickness. The technology of processing is therefore fundamentally significant for a tyre retreader [1].

In 1864, ink-maker J.K. Wright of Philadelphia (PA, USA) discovered the process of manufacturing carbon black. Improvements in production technology were subsequently developed and, in the early 1900s when the Goodrich Tyre Company started using carbon black, they found that it significantly improved some of the properties of tyre rubber compounds.

https://doi.org/10.1515/9783110643893-002

Carbon black became widely used after 1915 as a reinforcing agent for the production of automobile tyres, with the effect of increasing both durability and strength. Following the addition of 50 phr (per hundred of rubber) carbon black to rubber compounds, road wear resistance and tensile strength were improved to a great extent. Industrial uses of carbon black today are usually in the form of furnace black and thermal black.

The reinforcing material carbon black, one of the major ingredients in tread rubber compounds, has significant effects. Its impacts include high hardness, extrudability and cure rate. Its effects on mechanical properties such as tensile strength, modulus, tear strength and fatigue are considerable, but it has only a medium impact on resilience and ageing.

On exposure to direct sunlight for a long period a tyre may become increasingly grey. The tyre also deteriorates gradually due to the effect of ultraviolet (UV) light. Carbon black, when incorporated, absorbs UV rays and protects the tyre. An illustration of the different grades of carbon black utilised in tyre retread compounds are discussed in this chapter [2].

Intended for the preparation of rubber-based cement, a high-rubber-content compound mixed with a suitable adhesion promoter such as natural rosin, synthetic resin, swollen and dissolved in a suitable solvent oil, is used in the assembly process for tyre retreading. The specific viscosity is obtained on the way to prepare the vulcanisable cement employed in the construction of retreading. For the purpose of cement preparation, the base rubber needs to be suitably premasticated before incorporating other compounding ingredients in the mixing process.

Rubber cement is essential for retreading assembly. It is prepared in suitable solvent/s and applied with uniform thickness on the ready worn tyre surface to adhere perfectly to the used tyre. In the cold process for retread tyre moulding, as is usually used for car and truck tyres, gum cement preparation is required for fabrication of the precured tread on the used tyres.

Until the mid-1950s, the only antiozonant for rubber compounding available to rubber compounders was wax. Its limited solubility in rubber led to blooming at the surface, and it formed a protective layer that did not allow ozone to permeate, but in dynamic applications the use of wax is ineffective. In 1954, ozone protective agents for dynamic applications were introduced. It is very important to incorporate antiozonant/antidegradants such as piperidine pentamethylene dithiocarbamate and isopropyl-phenyl-*p*-phenylenediamine in tyre tread formulations in order to provide protection against oxygen and ozone attack as well as resistance to surface cracking.

Cracking of tread rubber during service is largely a result of the effects of oxygen on high flexing of the tyre. Ozone is particularly active in producing this detrimental effect. An antioxidant added into the tread rubber formulation in combination with a suitable antiozonant will safeguard the tread from attack from oxygen and ozone, resulting in the intensification of flexing, as discussed in this chapter [3].

Tyres may be attacked by microorganisms, fungi and some bacteria. These infiltrate the rubber and weaken its double bonds, in particular when the tyre is used in marsh/wetlands for long periods, ultimately damaging the tread. Accordingly, for such applications it is necessary to protect the tread compound by using special additives, as also discussed in this chapter [4].

2.2 Processability

Rubbers are highly viscous, so it is necessary to reduce the molecular weight (MW) and viscosity of elastomers to facilitate processing of rubber compounds. Decreasing the viscosity and increasing the plasticity of the elastomers used for tyre retreading applications can be achieved by mechanical milling, either using open two-roll rubber mills or internal mixers (the Banbury mill). This initial process is necessary to incorporate reinforcing fillers, PA and other powdery compounding ingredients into a homogeneous mix of the desired plasticity for use in retreading processes for various types of tyre of different designs.

Processability is critical to the properties required in compounding to produce an ultimate product. Each compounding ingredient should be dispersed in the rubber matrix consistently and it should have uniform rheological properties in order to provide the desired attributes in the end product. On preparation of the rubber mix, a highly viscous material behaves as a non-Newtonian liquid.

The initial mastication of natural rubber (NR) can be performed in an open two-roll rubber mill or in an internal mixer. Repetitive mechanical deformation in the presence of atmospheric oxygen causes a reduction in the MW of the rubber, and it becomes increasingly plastic. At this stage, rubber compounding ingredients can be incorporated with further mixing. Synthetic rubbers such as styrene-butadiene rubber (SBR), polybutadiene rubber (PBR), ethylene propylene diene monomer (EPDM) and chloroprene rubber (CR) can be treated like NR during the mixing operation [3].

To speed up the mastication process, chemical peptising agents such as pentachlorothiophenol (Renacit VII), together with dispersing additives and zinc-2- benzamidothiophenate (Pepton 65), in dosages of 0.2–0.5 phr will assist in breaking down rubbers at a lower operating temperature. Higher proportions may be necessary for synthetic rubbers. The mill temperature during mastication of NR should be maintained at about 70 °C, and for synthetic rubbers 130 °C may be necessary to reduce the mastication time effectively.

Lowering the viscosity with peptisation will facilitate the incorporation of a high proportion of fillers and other compounding ingredients during mixing. This can assist with flow properties during the extrusion of tread profile and in the calendering of tread/cushion gum strips, and all through the process of moulding. Appropriate peptisation of the rubber, particularly high-viscosity NR, will help in blending with other synthetic elastomers, which are available in different viscosity levels (even lower viscosities, e.g., general purpose synthetic rubbers).

The advantages of correctly peptising rubbers during compounding are as follows:

- Improvement in blending of elastomers
- Faster incorporation of fillers and other compounding ingredients
- Decreased processing temperature
- Facilitated processing of extrusion/calendering/moulding
- Increased building tack of tread and other components
- Reduced batch-to-batch variation of mixed stocks
- Reduced time of mixing
- Reduced cost

Plots depicting the effects of peptisation on NR (without and with peptiser) are presented in Figure 2.1 [5].

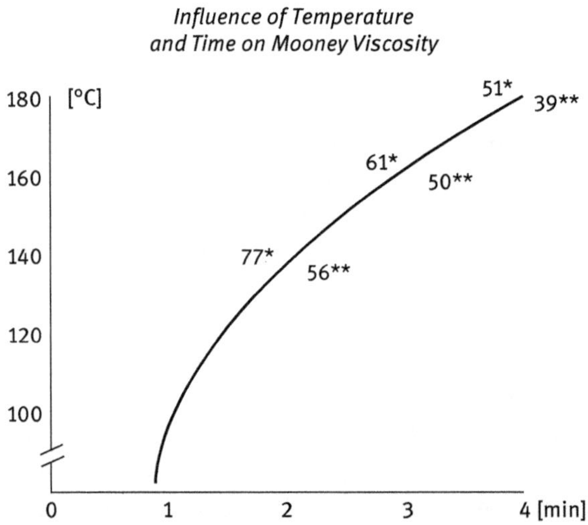

Figure 2.1: Peptisation of NR without* and with** peptiser (influence of temperature and time on Mooney viscosity). Reproduced with permission from *Rubber Handbook – Schill + Seilacher*, Struktol, Stow, OH, USA, 2004, p.30. ©Struktol and Schill + Seilacher [5].

The addition of homogenising agents (Struktol 40MS/60MS) into tread formulations will promote the homogeneity of elastomer blending, smoothen the incorporation of compounding constituents during the mixing process, and also help minimise batch-to-batch variations in mixed stocks. These agents are composed of a mixture of aromatic hydrocarbon resins and display good compatibility that facilitates the blending of rubbers. The usual effective dosage for NR is 4–5 phr, and this gives satisfactory results in general purpose synthetic rubber blends such as NR, SBR and PBR. For special purpose synthetic rubber blends such as CR or EPDM, a higher proportion of 7–8 phr is necessary to achieve optimum processing effects.

As a result of the excellent wetting properties of compatibilisers, fillers can be incorporated more quickly and dispersed more evenly during mixing. They also increase the green strength of compounds and improve green tack, which helps in the construction of retread as a result of superior homogeneity. Flow properties are also improved, which facilitates curing.

The following benefits arise incorporating homogenising agents into retread rubber compounds:

- Improved homogeneity of stocks/reduced nerve
- Better batch compactness
- Smoothness of compounds
- Improved processing (e.g., extrusion and calendering)
- Better dispersion of fillers in compounds
- Increased building tack
- Uniformity in processing behaviour

A plot of viscosity reduction *versus* temperature on peptisation of NR is shown in Figure 2.2.

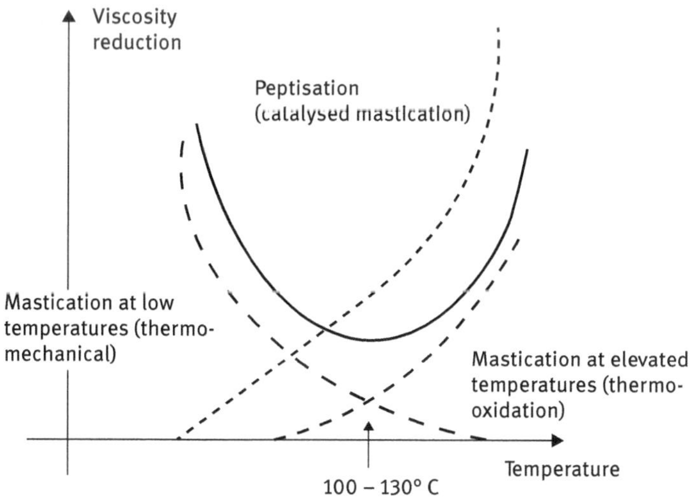

Figure 2.2: Peptisation of NR viscosity reduction *versus* temperature. Reproduced with permission from *Rubber Handbook – Schill + Seilacher*, Struktol, Stow, OH, USA, 2004, p.30. ©Struktol and Schill + Seilacher [5].

The incorporation of some PAs (e.g., petroleum-based process oils, plasticisers, resins, factices and zinc salts of unsaturated fatty-acids, internal lubricants, waxes, tackifying resins, and pine tars) considerably improves processing behaviour, including mill mixing and extrusion and calendering operations in natural/synthetic rubber-based stocks.

The utilisation of zinc soaps of higher fatty-acids in mixes may assist in avoiding the problem of sticking to the mill during mixing, and is effective in the extrusion of tread profile and calendering of tread/cushion gum sheeting. They can also aid in the dispersion of fillers, and will improve the flow behaviour of mixes while moulding the tyre, leading to a better finish of the retread [5].

To promote the tack of rubber compounds, resins such as coumarone–indene resins (CI resin), petrolium resins, phenolic resins and wood rosin are added, as well as to improve processing, incorporate dry fillers during mixing and facilitate extrusion/ calendering operations.

The essential results of these actions in the processing of rubber compounds for use in tyre retreading applications are:

- Appropriate viscosity (to improve flow properties)
- Adequate green strength on handling
- Control of shrinkage
- Appropriate tack (to impart adhesion with the old tyre substrate during assembly)

The benefits of using a silane coupling agent (Struktol SCA 98/SCA 985) in tread rubber compositions where a high proportion of reinforcing silica filler is incorporated arise from coupling of the silica fillers to the elastomer backbone, thus improving the reinforcing potential. The following properties of the tread compound at the time of processing and in cured tyre tread are improved:

- Rate of incorporation of fillers (faster)
- Dispersibility of ingredients (improved)
- Viscosity of the rubber compound (reduced)
- Modulus of the compound (enhanced)
- Abrasion resistance property (increased)
- Rolling resistance (improved)
- Hysteresis (lowered)

Petroleum oils of aromatic type with a low aniline point improve the processing of NR-based compounds, helping in the dispersion of powdered fillers and maintaining the viscosity of the mixes for further processing. These oils act as an extender to rubber mixes; aromatic oils are suitable extenders for NR and general purpose synthetic rubber blends, and naphthenic/paraffinic oils are suitable for EPDM blends with NR [1].

In terms of the effect of oil extender/plasticisers in retread rubber mixes, the following advantages can be achieved:
- Lower compound viscosity
- Faster incorporation of fillers
- Reduced generation of heat during mixing
- Improved flow properties during processing
- Increased building tack
- Lower compound hardness
- Greater elongation
- Enhanced flex life

Because NR has a high degree of unsaturation in its polymer chain, there is a tendency towards premature curing or 'scorching' of NR-based compound during processing in the extruder or calender. This processing problem is also observed for some synthetic rubber mixes when reinforcing carbon blacks are added in a high proportion to the rubber compounds, when accelerators are added in a high dosage or when processed at higher temperatures. The 'heat history' of processed mixes from the stage of mixing and including the temperature of storing the compounds influences this kind of phenomenon.

To extend the scorch time to a safe level for the processing temperature of rubber mixes, chemical rubber retarders such as cyclohexylthiopthalimide, nitrosodiphenylamine and benzoic acid are added in dosages of 0.1–0.3 phr. These can extend the level of operational safety while processing unvulcanised compounds [3].

2.3 Requirement of mechanical properties

Carbon black is used in rubber compounds to augment mechanical properties such as tensile strength, modulus, abrasion and tear resistance. Selection of the proper grade of carbon black to achieve these specific physical characteristics for tread applications is described in Table 2.1.

The physical properties are improved because the carbon black aggregates create physical and chemical bonds with the rubber that give the requisite tensile strength, tear strength, abrasion resistance and flex fatigue resistance [6].

The reinforcing filler precipitated silica provides higher reinforcement properties in polar elastomers like CR, and in NR/PBR/SBR blended stocks reinforcement may improve on using silane coupling agent along with the silica. Non-black reinforcing fillers can be included in compounding; for example, coupling agents in mixes improve coupling to the elastomer backbone. They can increase the modulus of the compound, improve abrasion resistance, give low hysteresis and improve rolling resistance.

Table 2.1: Types of carbon black used for tyre retreading formulations.

Type of carbon black	ASTM designation	Particle size (millimicron)	Surface area (m^2/gm)	Oil absorption (cm^3/g)	Dibutyl phthalate absorption number (cm^3/100 g)	Iodine absorption (g/kg)	Pour density (kg/m^3)	Heat loss (% maximum)
SAF	N110	20–25	110–120	1.50	113	145	370	3.0
ISAF	N220	24–33	100	1.35	114	121	344.43	3.0
HAF	N330	28–36	65–70	1.30	105	82	376.4	2.5
FEF	N550	40–55	40	1.35	121	43	360.4	1.5
SRF	N762	61–100	33	0.7	65	27	350	1.5
GPF	N660	49–60	36	1.1	90	36	440	1.0
MT	N990	250–350	10	–	43	6–12	640.8	0.10

ASTM: American Society for Testing and Materials FEF: Fast extrusion furnace
GPF: General purpose furnace
HAF: High abrasion furnace
ISAF: Intermediate Super Abrasion Furnace
MT: Medium thermal
SAF: Super Abrasion Furnace
SRF: Semi-reinforcing furnace

Fine-particle silica can also contribute to improving properties such as tear, cut and chip resistance, and also lowers the heat build-up characteristics during high tyre flexing conditions. An average silica surface area of about 152 g/m^2 is acceptable for these purposes [7].

NR exhibits good abrasion and tear resistance characteristics, and the incorporation of reinforcing carbon blacks into compounds of NR further improves these characteristics, as well as the tensile properties.

In comparison with NR, SBR exhibit better crack initiation and abrasion resistance characteristics than NR. PBR can provide excellent abrasion resistance when blended with NR in tyre tread compositions [8]. Because PBR shows higher abrasion resistance when blended with NR, the addition of some percentage of this rubber into mixes based on NR, SBR and CR can enhance this property noticeably. Because NR, being a non-polar rubber, can blend easily with other non-polar rubbers, it can contribute its excellent properties of tear and abrasion resistance to blends [9].

The small particle size and high surface area of carbon black helps to increase the mechanical properties of rubbers. To obtain extruded tread and calendered gum materials with dimensional stability and to achieve surface perfection, proper dispersion of blacks into mixes is essential.

The use of dual-phase fillers such as a blend of carbon black and silica in NR-based tread compounds for heavy-duty truck tyres and off-the-road (OTR) tyres on the whole can provide improved performance due to the high elastomer–filler interaction and low filler–filler interaction. In comparison with silica alone, the carbon–silica dual filler can contribute considerably to the improvement of tearing strength and abrasion resistance. Further improvement can be achieved by the addition of a small dosage of coupling agent into the tread formulation [e.g., *bis*(3-triethoxysilylpropyl tetrasulfane) (Struktol SCA 98] [10].

Some resins (e.g., coumarone resin, petroleum and phenolic resins), when added in the right proportion (about 4–5 phr), can contribute to improving tear resistance. Excellent levels of properties such as abrasion resistance are achievable for EPDM, similar to those for NR, when blended with NR and PBR.

Proper curing of the elastomer plays an essential role in determining abrasion resistance, and it may deteriorate if the crosslinking is not adequate [11].

2.4 Perceptions of reinforcement in rubber compounding

Mixes used for tyre retreading applications can include fillers of the reinforcing type to improve the properties of the end products as well as some inert fillers to reduce the cost of the compounds. Fillers improve the properties of the elastomer by enhancing mechanical properties such as abrasion resistance and fatigue life.

Different reinforcing fillers are incorporated into rubber compounds to enhance mechanical properties such as modulus, tensile strength and wear resistance. The most commonly used are carbon black and silica. Reinforcing fillers in retreading formulations are significant in enabling demanding performance requirements to be met, in particular in relation to resistance to wear. They impart a degree of strengthening to the rubber network, leading to considerable enhancement in tensile strength, modulus and resistance to abrasion, as well as an increase in endurance. Such improvements are significant because strength is improved throughout the tread, particularly in its footprint region, thereby improving road grip and decreasing the stopping distance of articulated vehicles, for example, and contributing to road safety for passenger vehicles [7].

Reinforcing fillers achieve these properties by ensuring a high number of chemical linkages within the elastomer network. Non-reinforcing fillers, however, do not form chemical links with the elastomer and so do not provide any considerable improvement in this regard.

While designing a product and analysing the service properties of a critical item such as tyre tread, it is imperative to understand the effect of filler in the rubber composite. In elastomer formulations the distribution of particle sizes is important, as this affects the mechanical properties and therefore the performance of the ultimate product [12].

Fillers impart distinctiveness to a rubber compound as a result of their particle characteristics, including size, surface area, surface activity and shape. A summary of strength and stiffness of SBR with and without reinforcing filler is shown in Figure 2.3.

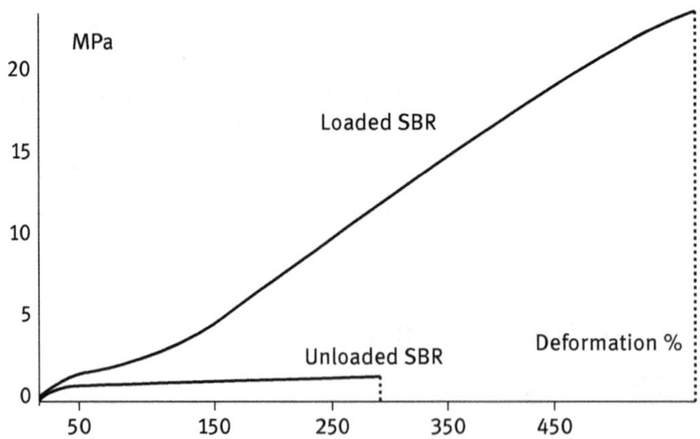

Figure 2.3: Strength and stiffness with and without reinforcing filler in SBR. Reproduced with permission from *Reinforcing Fillers in the Rubber Industry*, Fact Sheet, European Tyre and Rubber Manufacturers' Association (ETRMA), Brussels, Belgium. ©European Tyre and Rubber Manufacturers' Association [7].

Commercially available rubber grades of carbon black have particle sizes ranging from 10 to 500 nm. The larger the particle size, the smaller the reinforcement property. Surface activity affects the compatibility of the filler with the rubber and the ability of the rubber to integrate with the filler. Properties depend on the dispersibility of the carbon black in mixes, and this usually depends on the size of particle, hardness of the pellets and structure.

The reinforcement of rubber compounds, in particular the interaction of filler and rubber, should be examined in order to improve vulcanisate mechanical properties, including tensile strength, tear resistance, abrasion resistance and modulus. Incorporating fine-particle fillers into rubber mixes improves the physical characteristics of the cured product as a result of these rubber/filler interactions. Reinforcing fillers provide a large degree of strengthening of the rubber network by providing a high number of chemical links with the polymer network [13]. A graphical representation of the tensile strengths of different rubbers is presented in Chapter 1 (Figure 1.1).

The reinforcing fillers used in tread rubber compounds are essential to achieving the desired performance requirements. Tread compounds have demanding requirements regarding resistance to wear. They should also have the durability that allows their use at high-speeds over a long period, as well as the effect of lowering fuel consumption, as far as possible [14].

The following are essential factors that intensify the reinforcement of elastomer properties:
- Size of the particles or specific surface area (in combination with loading, determine the effective contact environment of the filler and polymer network)
- Structure of the filler particles, and to an extent their aggregation (affect the physical interaction between filler and rubber)
- Type of filler surface and intensity of the interaction between rubber and filler (contribute to restricting motion of the rubber molecular chain while under stress)

Thus, surface activity is a dominant factor in filler–filler and filler–rubber interactions [15]. The reinforcement of high abrasion furnace (HAF) carbon black per hundred of rubber (phr), and with aromatic process oil is illustrated in Figure 2.4.

2.4.1 Reinforcing carbon blacks

The physical properties of tensile strength, flexing, abrasion and tear resistance of NR are usually high. The incorporation of reinforcing carbon blacks, such as SAF, ISAF and HAF, in tread mixes will further enhance the tensile strength, abrasion and tear resistance of this elastomer. The characteristics of different carbon blacks used in tyre retreading compounding are displayed in Table 2.1.

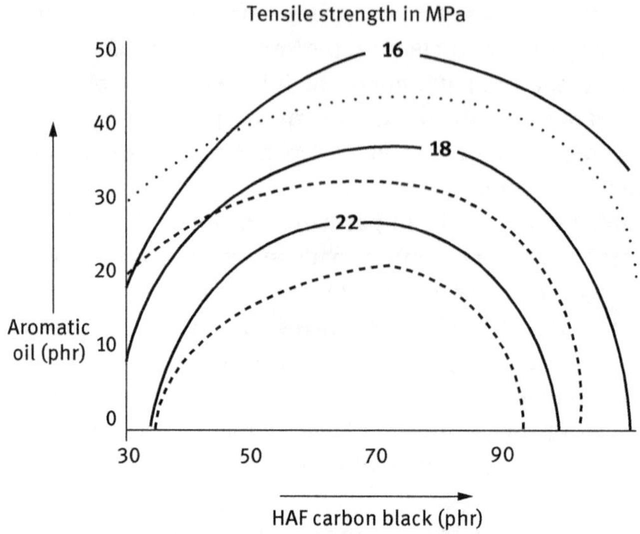

Figure 2.4: Reinforcement of HAF black with process oil.

In general, carbon blacks are graded according to their surface area parameter: (i) semi-reinforcing (<45 g/m^2) and (ii) reinforcing (65–140 g/m^2). Pneumatic tyre retread made from NR or using a blend of different rubbers is typically reinforced with carbon black as a major reinforcing filler [12]. The properties of different grades of carbon black in NR-based mixes are shown in Table 2.2.

Table 2.2: Properties of carbon black in NR.

Grade of carbon black	Hardness (Shore A)	Tensile strength (MPa)	Modulus at 300% (MPa)	Elongation at break (%)	Rebound resilience (%)
SAF	65	25.6	12	500	60
ISAF	61	25	12.2	520	60.5
HAF	60	24.8	11	510	65
FEF	58	22.4	13.5	480	70
SRF	57	21.8	11.8	490	71.6
GPF	55	19.5	11.5	530	72.5
MT	49	20	4.5	620	84

Base recipe: NR (RSS 1X), 100; zinc oxide, 5; stearic acid, 2; carbon black, 50; process oil, 3; cyclohexyl benzthiazole sulfenamide, 1; and sulfur, 2.5

For mixes of tread gum or cushion gum compounds, the preferred grade of carbon black is FEF, as this provides improved calendering properties compared with the next choice of SRF. Furnace blacks such as SRF/GPF grades are suitable variety for the compounds utilise in repair mixes and for rubber cement compositions [13].

2.5 Reinforcing silica, silicates and carbonates for tyre retread

Silicas such as fumed silica and precipitated silica enhance the tear strength of rubber compounds due to their small particle size and complex aggregate structure. To achieve the highest degree of reinforcement in tread mixes, it is essential to select the appropriate grade and proportion. Precipitated silica is added to compound formulations together with carbon black to improve properties such as flex cracking and cut growth resistance.

Major improvements in rubber properties are achievable by using reinforcing fillers with particle sizes of 10–100 nm in rubber mixes. Equitable improvements in physical properties can be obtained with fillers such as calcium and aluminium silicates and activated/precipitated calcium carbonate. These partially reinforcing fillers can be used in tread gums, repair compounds, cushion gums or in cement recipes [16].

Reinforcing fumed and precipitated silica can enhance the following characteristics:

- Tensile strength (increased)
- Durability
- Abrasion resistance (enhanced)
- Tear resistance (enhanced)

The morphology of silica has an effect on its reinforcing properties with rubber. The following structural characteristics are key:

- The primary particles have cross-sectional dimensions of 5–100 nm.
- Aggregates (dimensions of 100–500 nm) are formed by chemical and physicochemical interactions.
- Agglomerates (typical dimensions of 1–40 μm) are fragmented during mixing of rubber compounds until about the size of aggregates or even primary particles.

Silica aggregates have a higher structure than carbon blacks. These structures are the reason for their better reinforcing power when compared to carbon black. Because of their high specific surface energy they have a strong propensity to agglomerate and are difficult to disperse in rubber. They also re-agglomerate rapidly after mixing [17]. The properties of silica, silicates and calcium carbonate are illustrated in Table 2.3.

Table 2.3: Properties of silicas, silicates and calcium carbonate.

Types of silica	Specific gravity	Average particle size (nm)	Surface area (g/m^2)	pH	Loss on ignition (%)	Oil absorption (g/100 g)
Fine fumed silica (anhydrous)	2.1	8	300–380	4	1.0	200
Precipitated silica (hydrated)	1.95	16	170	5.7	6	240
Calcium silicate	2.10	85	35	10	7	110
Aluminium silicate	2.10	25	130	10	5	140
Precipitated calcium carbonate	2.42	50	74	–	–	–

2.5.1 Specialty ingredients for tyre tread

The incorporation of short aramid fibres as filler in rubber retread compounds, particularly for use in OTR tyres, can offer processing advantages and improved strength at lower cost. Rubber compounds with the correct quantity of short aramid fibres will have high modulus and better tear strength. The amount of reinforcement depends on the kind of fibre, rubber matrix–fibre adhesion, and distribution of fibre length in the fibre filler. Also important is the mechanical contact between rubber matrix and fibres due to surface roughness.

The essential factor for achieving reinforcement using short fibres is adhesion between the fibre and elastomeric matrix. Improvements in adhesion can significantly increase the tensile strength and elongation at break of the compound. Increased adhesion with the elastomeric matrix is achieved by treating the fibre surface using resorcinol formaldehyde latex.

On adding short and cut lengths of aramid fibres to a mix, a substantial amount of fibre breakage takes place during mixing. However, sufficient fibre lengths are available in the stock to provide an adequate reinforcing effect, in particular low elongations in the end product [18].

2.6 Dispersion effect of fillers in rubber mixes

In order to achieve satisfactory mechanical properties in tyre tread compounds, a high degree of dispersion of fillers and other ingredients is essential. Inadequate dispersion of fillers (e.g., reinforcing fillers) will adversely affect basic properties such as tensile strength and abrasion resistance. Reasons for failure of tyre retread while in service include cracking due to fatigue caused by agglomeration of fillers in the mixes. Proper dispersion in the rubber matrix is therefore ensured by process testing of the tread compound. As well as tensile strength and resistance to abrasion, other critical properties also impacted include cracking on flexing and fatigue during service if agglomerates of fillers are present in mixes [14].

2.7 Adhesion to used/Worn-out tyres

The strength of bonding during adhesion of rubber tread, tread gum and cushion gum to an old tyre prepared for retreading depends on the nature of the rubber, its intrinsic tackiness and appropriate preparation of the used tyre. The process should not diminish the flexing endurance of the old tyre during elevated dynamic stresses while in service.

Fillers and a suitable tackifier provide an increase in adhesion level by enhancing the modulus and tear strength of the compound. A high modulus in a vulcanised retread matrix can reduce the difference in extensibility between the substrate and rubber cover, thereby reducing strain at the interface. A higher tear strength prevents the vulcanised rubber from being ripped from the substrate, and prevents chipping and cutting of the tread.

Improvement in properties like the modulus and tear strength of the compound, together with an increase in adhesion strength, can be achieved by incorporating reinforcing silica fillers. A higher level of tensile strength, modulus and tear strength can also be attained by incorporating reinforcing carbon blacks into the stocks. High-tensile-strength mixes impart relatively better adhesion properties with the substrate [19].

To promote the strength of bonding with the old tyre, some special-category additives may be incorporated into mixes, including reactive resins and a resorcinol – formaldehyde silica combination. These are in the form of dry bonding agents and are ready for use.

The addition of pine tar, wood rosins, coumarone–indene resins and other hydrocarbon resins can provide adequate tack in the uncured state. These help in the fabrication/ building-up of the tyres.

The advantage of adding resins into mixes is an increase in the adhesion strength by spreading the stress over a wider surface area, thereby promoting adherence with the substrate by mechanical action. During the process of vulcanisation, crosslinking takes place between the rubber and the resin, resulting in an increase in bond strength between the applied rubber compound and old tyre substrate [17].

Curing agents have a distinct effect on the bonding of old tyre and retread, and a reduction of the sulfur dosage in the compound may reduce the strength of adhesion considerably where a sulfur cure system is appropriate.

Correctly compounded unsaturated hydrocarbon rubbers such as NR and SBR demonstrate adequate adhesion in retreading. The appropriate choice of fillers (particularly reinforcing fillers, in the correct proportions) and tackifying agents gives good results.

Satisfactory adhesion is achievable by incorporating unsaturated hydrocarbons PBR into mixes. Marginally superior adhesion strength is achieved with SBR compared to NR-based stocks.

When polar backbone structure elastomers like CR are blended with NR in high proportions and used in retreads for special applications of tyres, very good bonding with old substrates is seen. Highly saturated elastomers like EPDM have no intrinsic tack for good adherence with the used tyre; they need liberal amounts of tackifier in mixes and are usually blended with NR to build tack [20].

The attributes accountable for adhesion between tread and the old tyre are as follows:
- Mechanical adhesion (rough surfacing)
- Primary and secondary chemical bond formation
- Diffusion of adhesive/cement components

2.8 Properties of tear and abrasion resistance

The property of tear resistance is resistance to growth in cuts while under tension. Resistance to rubbing during service is termed the abrasion resistance of the rubber. A laboratory test of the abrasion resistance of a specimen may not be relevant to the actual service life of a retreaded tyre, because it is difficult to predict the actual use of a tyre when in service.

The resistance to abrasion of a compounded rubber for use in tread can be poor because of insufficient crosslinking or vulcanisation, but also deteriorates with excessive crosslinking. The optimum value of tear resistance can be achieved with a somewhat reduced degree of cure, and can decrease rapidly with over curing. The tear resistance will also decrease with increasing temperature during service.

In NR/PBR/SBR and polychloroprene rubber blended compounds, the incorporation of reinforcing fillers with reduced particle size (e.g., HAF, ISAF and SAF grades of carbon black) gives higher abrasion resistance. Precipitated/fumed silica gives higher abrasion resistance properties to rubber tread than other active fillers. The use of different grades of carbon black in tyre tread formulations to achieve high abrasion and high tear properties is detailed above (Tables 2.1 and 2.2) [21].

2.9 Resilience and flexibility

The property of resiliency of rubber to heat build-up is specifically defined as the ratio of return to resulting energy, and the heat generated is the measurement of the amount of energy absorbed. This property can be improved by increasing the level of crosslink density during vulcanisation, but declines with excessive crosslinking. With a higher level of crosslinking, rebound resilience accentuate raise of temperature, while at slightly excess in crosslinking exhibits a minimum level. During cyclic deformation, the heat build-up usually depends on the chemical structure of the polymer and the type of formation of crosslinking during curing.

Among the rubbers used for retreading, polybutadienes provide the best values of resilience, while NR, SBR and CR show high rates of resilience and flexibility [22].

2.10 Resistance to oxygen, ozone and flexing

In the course of service a tyre tread is stretched under dynamic conditions, and cracking can appear on the surface. The presence of ozone, even at very low concentrations, affects the rubber, and the extent of resistance to ozone attack depends on the nature of the elastomer. NR and SBR have the lowest resistance to ozone, and EPDM the highest.

The ozone resistance of elastomers can be increased by adding rubber antiozonants such as *p*-phenylenediamine and waxes into rubber compounds. Waxes may bloom to the surface because of their limited compatibility with rubber and offer a protective layer, but the surface layer may crack under high flexing conditions and will not therefore fulfil its purpose. When there is a tendency towards blooming on the surface, adhesion may be affected, so microcrystalline waxes are used instead and can provide the requisite effect [23].

The deterioration of rubber is triggered by oxygen, ozone, heat, light, metal catalysis and flexing, which affect the unsaturated part of the molecule that is accessible for sulfur vulcanisation. Two main classes of chemicals can impede the degradation process of unsaturated rubbers: amines and phenolics.

Antioxidants such as polymerised trimethyl dihydroquinoline in NR and general purpose synthetic rubber blend compounds provide protection against oxidation and heat under severe conditions. Resistance against flex cracking, heat and oxidation can be aided by adding diphenyl-*p*-phenylene diamine into rubber tread compounds [24].

2.11 Resistance to decay of tyre treads from natural causes

The deterioration of rubber automobile tyres known as 'dry rot' – a process of degradation that is accelerated by exposure to heat, ozone or UV radiation – may decay rubber tyres. This degradation can be indicated by web-like cracks in the tyre, which become larger and lead to rupture, resulting in the ultimate failure of the tyre. Proper care of tyres is essential to prevent dry rot. It can badly damage them and cause them to be dangerous during service. Dry rot is a kind of fungus that grows on the cellulose present in wood, and also depends on the proximity to dampness of soil bacteria. That dry rot does not usually grow on rubber, but a similar type of dry rot may appear on tyres in the same manner as this wood dry rot.

Intense exposure to UV rays may damage tyres, so tyres are manufactured with the addition of carbon black, which absorbs these destructive rays. Other protective agents such as the compounding ingredients antiozonants also protect the tyre from such damage.

The deterioration of rubber tread due to flexing while exposed to atmospheric ozone may lead to its breakdown. The addition of protective waxes such as microcrystalline waxes and in combination of rubber antiozonants will protect it from this impairment.

NR-based tyre treads may be prone to fungal attack, particularly when the retread compounds in the unvulcanised condition are stored for long periods in a damp/ humid atmosphere. The same can also happen to a cured retread tyre when exposed to a wet, drenched environment for prolonged periods while in use. Synthetic rubber-based treads are less prone to this problem compared to general purpose synthetic rubbers and NR-blend treads.

The use of antimicrobial chemicals in rubber compounds assists in preventing vulnerability to the attack of microorganisms, fungi and other microbes that may affect the service durability of tread. A mixture of phenols and sulfur compounds and dihydroxy-dichlorodiphenyl methane are available in powder form, and may be added in small dosages to stocks where such phenomena are prevalent. Infestation by microorganisms, and drying and cracking due to exposure to UV rays, oxidation and ozone, weaken the double bonds of rubber and carbonyl and aldehyde compounds, ultimately leading to deterioration of the rubber tyre tread [4].

References

1. B. Banerjee and S.K. Pal in *Retreading of Tyres – Methodology and Its Significance in Economics and Environment*, Book of Abstracts, India Rubber Expo – Conference and Workshop, Mumbai, India, 2013, p.189.

2. http://www.continentalcarbon.com/.
3. M.S. Evans in Tyre Compounding for Improved Performance, Rapra Review Report No.140, Rapra Technology Ltd, Shawbury, UK, 2001.
4. C. Smith in *How to Protect Car Tires from Rotting*. http://www.ehow.co.uk.
5. Schill and Seilacher, *Rubber Handbook*, Struktol, Stow, OH, USA, 2004, p.30.
6. W. Hofmann in *Rubber Technology Handbook*, Hanser Publishers, Munich, Germany, 1989.
7. *Reinforcing Fillers in the Rubber Industry – Fact Sheet*, European Tyre & Rubber Manufacturers' Association (ETRMA), Brussels, Belgium, 2010.
8. W. Kleemann and K. Weber in *Elastomer Processing*, Hanser Publishers, Munich, Germany, 1998.
9. *Manual for the Rubber Industry*, Farbenfabriken Bayer AG, Leverkusen, Germany, 1972.
10. *Developments in Rubber Technology – 4*, Eds., A. Whelan and K.S. Lee, Elsevier Applied Science, London, UK, 1987.
11. *BS 903: Methods of Testing Vulcanized Rubber*, British Standards Institution, London, UK, 1950.
12. *ASTM D1765–13: Standard Classification System for Carbon Blacks Used in Rubber Products*, American Society for Testing and Materials (ASTM), Conshohocken, PA, USA.
13. V. Jha in *Carbon Black Filler Reinforcement of Elastomers*, Department of Materials, Queen Mary, University of London, London, UK, 2008.
14. R. Brown in *Physical Testing of Rubber*, 4th Edition, Springer Science & Business Media, New York, NY, USA, 2006.
15. *Rubber Technology*, Ed., J.S. Dick, Hanser Publishers, Munich, Germany, 2001.
16. *Science and Technology of Rubber*, 3rd Edition, Eds., J.E. Mark, B. Erman and F.R. Eirich, Elsevier/Academic Press, London, UK, 2005.
17. *Engineering with Rubber*, Ed., A.N. Gent, Hanser Publishers, Munich, Germany, 1992.
18. M. Shirazi and J.W.M. Noodermeer in *Proceedings of the 9th Fall Rubber Colloquium (KHK)*, 3–5th November, Hanover, Germany, Elastomer Technology and Engineering Department, University of Twente, Netherlands, 2010.
19. *Handbook of Rubber Bonding*, Ed., B. Crowther, Rapra Technology Ltd, Shawbury, UK, 2001.
20. B. Banerjee in *Proceedings of the National Conference and Exposition on Rubber and Allied Materials*, Indian Rubber Institute, Karnataka and Rajasthan Branch, Mysore, India, 2004, p.304.
21. *Developments in Rubber and Rubber Composites – 1*, Ed., C.W. Evans, Applied Science Publishers Ltd, London, UK, 1980
22. Cabot. http://www.cabotcorp.com.
23. F.W. Barlow in *Rubber Compounding*, 2nd Edition, Marcel Dekker, New York, NY, USA, 1993.
24. *The Vanderbilt Rubber Handbook*, R.T. Vanderbilt Company, Norwalk, CT, USA, 1978.

Bireswar Banerjee

3 Manufacturing technology for tyre retreading

3.1 Introduction

Retreading is a technological process in which a used tyre's worn tread is removed and a new tread is positioned on the undamaged tyre casing. It is the only option considered as recycling for the purpose of fresh utilisation of a tyre that has become unusable. Vehicle tyres constitute one of the major operating expenses of most fleets, but, with retreading, a tyre can provide the same mileage as a new tyre, perhaps even at a lower cost per kilometre. Indeed, an investigation into the causes of failure of tyres in Russia found that 70–90% of tyres fail due to tread wear during the course of service.

Quality off-the-road (OTR) tyres are designed to be retreaded, because if a new tyre is discarded after its tread has worn-out, the user may only have consumed 50% of its value. Retread OTR tyres cost only half as much as new OTR tyres, so retreading represents a substantial cost saving that does not compromise on quality. Studies have revealed that 15% of total fleet operating costs are related to tyres, and there is the scope to reduce this by up to 50% by retreading at the proper time, thereby saving 8–15% of operational expenses. Furthermore, retreading helps in overcoming the global shortage of new OTR tyres. From an economic point of view, retread truck tyres lead to savings of over $3 billion per annum in North America alone for truck operating companies.

Environmental impact is also a consideration. A superior quality retread OTR tyre requires only 40 kg of oil to retread, but about 200 kg of oil is required to produce a new tyre. Retreading can therefore save millions of kilograms of oil annually worldwide, thereby helping to conserve natural resources [1].

To guarantee the safety of retread truck tyres they are manufactured according to rigorous industry standards set by the UK and US Department of Transportation federal safety standards for the manufacture of retreaded passenger car tyres. As the parameter of identification, all retread tyres are marked with the manufacturer's name and date of production. Accordingly, retread tyres can be used safely in daily operations in aircraft, passenger cars, trucks, buses, military vehicles, ambulances and so on. Retreaded tyres can be driven at the same permissible speeds as comparable new tyres with no reduction in safety, performance or comfort for passengers.

From economic, operational and environmental aspects it is important to retread a tyre at the right time. Globally, about 80% of airline, off road and heavy-duty trucks (including lighter vehicles) are operating with retreaded tyres.

https://doi.org/10.1515/9783110643893-003

In terms of fuel economy, tyres play an essential role, as selecting the right tyre and utilising an appropriate tyre management system can save 4% fuel in operation. If the tyre is under-inflated it has more rolling resistance because of disproportionate tread distortion. Indeed, high rolling resistance and under-inflation can increase fuel consumption by up to 30%. Total tyre logic, or a proper tyre management system, is therefore as important to fuel economy as tyre selection.

Generally, retreaded tyres offer more rolling resistance than original treads, so the design of the tread pattern affects fuel economy. A tread design with lower depth pattern can offer better fuel economy, although the effect may vary in different operations.

Tyre retreading industries, with their aim to repair, retread and recycle, are considered to be green and have become a national and international focus. They have the potential to save tons of steel and millions of barrels of oil, and reduce the emissions of carbon, thus offering colossal environmental benefits.

Before preparing a used/old tyre for retreading it is imperative to carry out a careful inspection for its suitability for retreading to ensure its service performance afterwards. In the 1980s, a non-destructive way of inspecting worn tyres was introduced; details of this inspection process are discussed in this chapter [2].

3.2 Retreading methodology

Retreading of tyres is as old as the manufacture of tyres. Indeed, it is a well recognised and reliable technology. Millions of such tyres are used for aircraft, trucks and OTR heavy duty vehicles, and cars are retreaded every year and sold worldwide to reduce fleet operating costs. Nearly 80% of the world's airline and heavy-duty vehicles use retreaded tyres.

Retreads are safe and can save hundreds of millions of gallons of oil compared to the levels consumed in the production of a new tyre. A study has revealed that 70–90% of tyre failure is caused by wear. To increase the life of tyres it is therefore necessary to increase the wear resistance of the tread, which can be achieved at the time of retreading. Retreading is a process that provides tyres with a second life by giving new tread to worn tyre [3].

The desirable basic requirements of a tyre are as follows:

- High wear resistance and long tyre life
- High wet grip to offer high security
- Low rolling loss to lower fuel consumption

The following are essential requirements for the development of retread rubber compounds:
- Processability
- Good green tack and green strength
- Low hysteresis
- High skid resistance
- High abrasion resistance
- Low cut growth rate
- Rolling efficiency
- Thermo-oxidative stability
- Economical

The operations for the production of a retread tyre are carried out in the following order:
1. Initial inspection of worn tyre
   The object of the preliminary inspection is to confirm that a casing is acceptable for retreading. This entails a thorough inspection of each tyre, inside and outside, identification of any casing impairment, and assessment of its reparability. Approve the casing for further processing, if acceptable within the criteria of inspection.

The next steps are important parts of the visual inspection procedure that is performed in order to accept a casing for retreading:
2. No separation of ply
3. Bead wire is not broken, damaged or exposed
4. No damage to plies in the bead area
5. No ply cords are loose and no sign of overload, under-inflation or having been run-flat
6. No parting of tread that cannot be removed during buffing
7. No separation in sidewall area
8. No cracking extending into body plies due to ageing
9. No non-reparable damage to the inner liner or bead area of a tubeless tyre
10. No large damage that cannot be repaired
11. No belt separation in radial tyres

Every tyre, while in use, is subject to wear and tear, some to a lesser degree than others. Regardless of the level of wear, providing the casing of the tyre is undamaged, which is so in about 75% of tyres, retreading is achievable. The tyre can recuperate its original condition and is prepared for use.

Millions of tyres are removed from vehicles every year and their usefulness restored by way of precise retreading. Accordingly, improvements in retreading are important, and end-of-life tyres have been studied in great detail.

3.2.1 Non-destructive tyre testing systems

The inspection of worn tyres by means of a non-destructive methodology using machines can give a better-quality retread *via* a process of selecting the right tyre before retreading. An X-ray device is available to the retreading industry, as well as shearography, a new and advanced holography method, both in the category of non-destructive equipment. These are used in combination with conventional visual inspection to give a rapid production process and accurate testing that includes:
- Electronic device
- Ultrasonic equipment
- X-ray machine
- Electro-mechanical system
- Holography
- Shearography or other types of casing inspection equipment that can aid in determining casing integrity

3.2.1.1 Shearography

The shearography machine is an important device that can accurately detect blemishes in the form of bubble flaw in tyre tread, shoulder and sidewalls. This system maps the surface of the casing using laser holography, and is an advanced non-destructive tyre casing examination process. During testing using a phase shearography system, a vacuum pressure difference of about 50 mbar is created that highlights any deformity caused by the expansion of air pockets. The equipment generates a holographic image of the damaged area/void and can calculate its dimensions and position, thus automatically assessing the tyre for further processing. This testing can ensure the durability of the casing after retreading. A schematic diagram of a shearography testing machine in operation is shown in Figure 3.1.

3.2.1.2 Laser shearography

Usually, worn tyres are inspected manually by striking using a hand tool and listening to the sound. The laser shearography functions like an X-ray to scan for any impairment or delamination of the tyres.

3.2.1.3 X-ray inspection

With this equipment, all internal components of the tyre can be viewed, including any void, air pocket or foreign material. This optical inspection method can provide

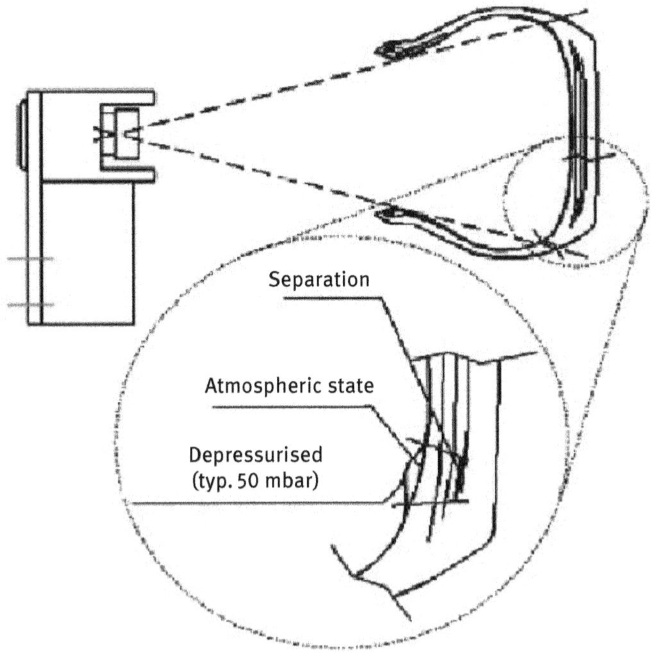

Figure 3.1: Shearography equipment for tyre testing. Reproduced with permission from Desser Tire & Rubber Co., Inc., Montebello, CA, USA.

adequate information about the internal structure of the worn tyre and its suitability for retreading for further processing. A high level of quality control can be facilitated by implementing a digital X-ray inspection system, and, because of its speed, it can be directly incorporated into the production process in a retreading plant.

The advanced X-ray inspection system is a superior microbeam radiation therapy process, which can expose the condition of the casing and the region containing the cables inside the tyre casing. It can also show the condition of the steel belts and cables concealed within the casing. This system of inspection can detect road hazard damage, ascertain any possible damage, and assess the condition of any earlier repair, but casings are not all required to undergo this procedure. An X-ray inspection system is depicted in Figure 3.2.

3.2.1.4 Holography

Holography or shearography tests are non-destructive inspection systems. A holography system can calibrate bubbles against a standard bubble, provide a holographic image of the bubbles, and can measure and provide the location of a bubble. On this basis it can grade a tyre according to the dimensions and sites of pockets.

Figure 3.2: X-ray testing of a tyre.

A hologram is important in that it can identify all the internal flaws of a tyre before selecting it for retreading, and after finishing the retreading can carry out bead to bead inspection. This device is effective in obtaining images of a tyre, and by inspection through cathode ray tube screen presentation and video data storage detects very small flaws that affect the functioning of the tyre. It can determine the ultimate acceptability of a tyre for retreading. An image of holographic testing equipment is shown in Figure 3.3.

Figure 3.3: Holography testing of a worn tyre.

3.2.2 Manufacturing process of retreading

During the retreading operation, the processing stages are in the following order:

(i) buffing, (ii) cementing, (iii) building and (iv) moulding. All steps are performed under adequate lighting to detect minor variations, and are essential for the retreading process in all categories of tyres [4].

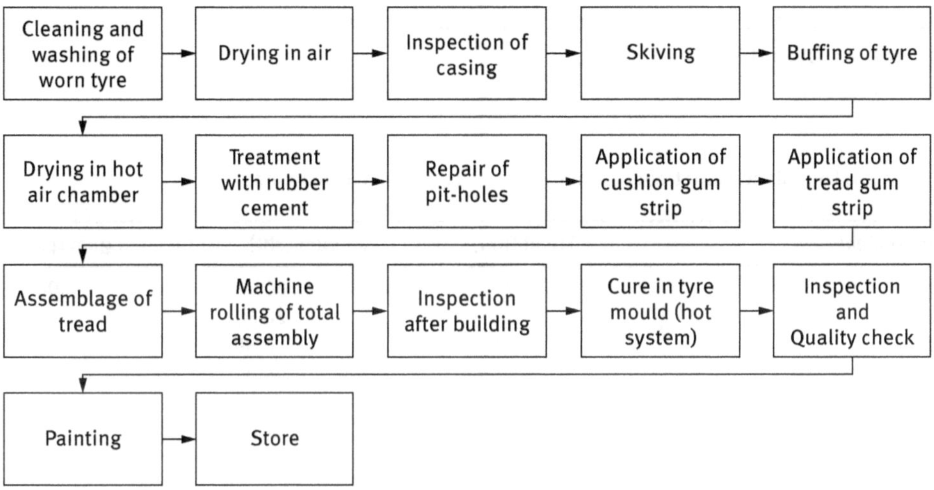

Figure 3.4: Manufacturing flow chart for hot tyre retreading system.

The steps undertaken to retread a worn tyre are as follows (Figure 3.4):
- Initial inspection of tyre
- Cleaning of worn tyre
- Skiving of old tread
- Whole tread buffing on machine
- Pithole buffing by hand grinding machine
- Repairing of pitholes
- Cementing
- Application of cushion gum
- Application of tread gum strip
- Laying of tread
- Pressure rolling on machine
- Moulding and finishing

- Painting of finished tyre
- Inspection and testing

The process of manufacturing tyre retread also makes use of a hot curing system.

3.2.3 Formulation for tyre retreading compounds

The model formulations for tyre retreading rubber compounds used in the manufacture of tread, tread gum, cushion gum, repair and cement preparations are detailed in Tables 3.1 and 3.2 [5, 6].

Table 3.1: Formulations for tread, tread gum and cushion gum compounds.

Ingredients	Tread (phr)	Tread gum (phr)	Cushion gum (phr)
NR	75	80	100
PBR	25	20	–
Peptiser	0.25	0.30	0.30
Zinc oxide	5.0	5.0	5.0
Stearic acid	1.0	1.0	1.0
Coumarone – indene resin	4.0	5.0	5.0
Phenol formaldehyde resin	2.0	3.0	3.0
Intermediate super abrasion furnace (N 220) carbon black	35	–	–
High abrasion furnace (N330) carbon black	–	30	15
Precipitated silica	20	20	25
Struktol SCA 98	6.0	4.0	3.0
Struktol HT 202	5.0	4.0	4.0
Pine tar	3.0	5.0	5.0
Aromatic oil	5.0	–	–
TMQ	1.5	1.5	1.0
p-Phenylenediamine	1.25	–	–
BSM	1.2	1.25	1.30
Insoluble sulfur	2.5	2.5	2.5
Accitard RE	0.20	0.15	0.15

Accitard RE: *N*-cyclohexylthiophthalimide
BSM: Benzthiazyl-sulphenmorpholide
NR: Natural rubber
PBR: Polybutadiene rubber
Struktol HT 202: Blend of zinc soaps and fatty-acid esters
TMQ: Trimethylquinoline

Table 3.2: Formulation for tyre repair and rubber cement compounds.

Ingredients	Repair compound (phr)	Cement compound (phr)
NR	100	100
Activated benzamidophenyl disulfide (Pepton 44)	0.25	0.30
Zinc oxide	5.0	10.0
Stearic acid	1.0	1.0
Coumarone – indene resin	5.0	6.0
Phenol formaldehyde resin	3.0	4.0
Semi-reinforcing furnace (N770) carbon black	25.0	–
General-purpose furnace (N660) carbon black	–	10.0
Precipitated silica	20.0	15.0
Struktol HT 202	3.0	–
Pine tar	3.0	5.0
Aromatic oil	5.0	–
TMQ	1.5	1.5
BSM	1.2	1.25
Insoluble sulfur	3.0	3.0

3.3 Essential physical properties in designing retread rubber compounds

The magnitudes of the physical properties required of a tyre retread compound predict the service life of the retread. The following important mechanical properties should be considered to determine wear performance and cut resistance:

- *Specific gravity*: Retread rubber compounds should have a lower specific gravity to achieve higher length of tread profile per kg of mix. When the tread rubber compound is made exclusively of natural rubber (NR) the specific gravity of the compound is on the low side. Using a blend of NR and general purpose synthetic rubbers such as polybutadiene rubber (PBR)/styrene-butadiene rubber (SBR) gives improved abrasion and cut resistance of the tread compounds, while the addition of more reinforcing fillers leads to superior physicals, and possibly a marginal increase in specific gravity.

A higher specific gravity indicates the incorporation of mineral fillers into the compounds, including clays; addition of a high quantity of inert fillers has a detrimental effect on the tyre's physical properties, lowering the quality of the retread.

- *Hardness*: The hardness of a tread rubber compound is measured to give an approximate idea of its primary physical properties, its stiffness in terms of its load-carrying capacity, and also the amount of filler in the compound. It can also indicate the degree of crosslink density (the greater the crosslink density, the harder the vulcanised product).
- *Tensile strength*: This is the resistance to breaking under tension. The minimum tensile strength of a retread rubber compound is around 20 MPa. Tread rubber compounds, which require high crack and cut growth resistance should have high tensile strength, particularly in heavy-duty truck and OTR retreads such as those for earth mover/dumper tyres. Hence, tensile strength can act as a basic indicator of the quality of the rubber compound for use as a high performance tread.
- *Modulus*: The modulus of a rubber compound is a measure of the resistance to being stretched. This property has relevance to tread wear and is also implicated in the heat build-up properties of a tyre.
- *Resilience*: A resilient rubber compound quickly recovers from deformation, and thus generates less heat when under highly flexing condition in operation. A higher heat build-up will quickly lower the quality of the retread. It is therefore appropriate to use a resilient rubber compound for retread applications.
- *Elongation at break*: This is the ability of the rubber compound at break or rupture, and describes its rubbery nature during ageing.
- *Abrasion resistance*: This property has a great relevance to tyre tread, being the resistance to wear on dynamic contact on the road surface while running. It is an essential property for tread rubber as a measurement of reinforcement of the rubber composition in the manufacture of the tread.

The specifications of the desired physical properties of a standard tyre retread compound, contingent on the type of tyre to be retreaded, and its purpose, are given in Table 3.3 [5, 7].

Table 3.3: Physical properties of rubber retread compound.

Properties	Requirement
Specific gravity	1.12–1.14
Hardness	65–70
Tensile strength (MPa)	18–20
Modulus at 300% (MPa)	9.8–10
Elongation at break (%)	500–600

3.3.1 Technology of adhesion on retreading

The adhesion of new tread to an old tyre requires the consideration of a wide range of variables that have an effect on rubber-to-rubber bonding. Among others, these include (i) the chemical composition of the rubbers, (ii) their compatibility/incompatibility factors, (iii) their molecular weights and distributions, (iv) the additives used, (v) their crystalline and amorphous contents, (vi) their surface nature and chemistry, and (vii) their crosslink densities.

A retread rubber compound usually consists of a blend of NR and a general purpose synthetic rubber such as PBR/SBR. The primary requirement is to achieve improved adhesion of the rubber compounds, whether using cement compound, cushion gum or tread gum mixes, and ultimately the tread composition in the complete assembly on the worn tyre.

The addition of tackifying wood rosin and synthetic resins in compounding facilitates the assembly of different categories of rubber compositions to build a complete retread. The incorporation of the dry rubber bonding agent resorcinol/formaldehyde and silica into formulations will provide significant bondability. The dynamics of curing the complete retread assembly result in an excellent bond strength with the old tyre.

In the use of a resin system to increase green tack to facilitate build-up assembly and to obtain adequate adhesion, the resin should be added at the initial phase of mixing with the elastomers.

Dry bonding agents are used extensively in tyre manufacturing and, when added directly to rubber mixes, provide a high level of adhesion with the worn tyre after curing. The strength of bondability is retained on heating, in humid atmospheres and during ageing.

Resorcinol resin, in combination with hexamethylene tetramine (HMT) – a methylene donor – is used in rubber mixes to obtain superior bonding properties. The HMT decomposes while heating during curing, producing ammonia and formaldehyde, which react with the resorcinol to produce an enduring, highly crosslinked elastomeric network. Substituting HMT with a modified melamine compound such as hexamethoxymethylmelamine with silica provides a widely used bonding system for retreading. The modified resorcinol/formaldehyde resins are a new generation of resins, and when added in precipitated silica filled retread mixes lead to reduce rolling resistance and help in controlling fuel consumption.

During rubber mixing, the methylene donor is added in the final stage of mixing (with the curatives) while keeping the mixing temperature low so as to achieve processing safety by avoiding premature crosslinking (scorching), in order to obtain improved bonding properties while moulding.

In the course of achieving bonding of the unvulcanised tread and other rubber components used for retreading, compounding ingredients such as sulfur and antioxidants are added but can lead to blooming and create a critical barrier to long storage. This problem can be prevented by using insoluble sulfur in the compound, which is transformed into its soluble form at about 90 °C. In the process of building with retread rubber components such as cushion gum, tread gum and the retread profile, the practice of solvent wiping using naphtha will help to remove bloomed ingredients from the surface before the assembly of the next components. Naphtha quickly evaporates from the surface, leaving a tack-free surface texture that assists in increasing the mobility of the elastomeric chain on removing the blooming chemicals [8].

3.4 Steps for manufacturing retreading

3.4.1 Buffing

Tyres must be buffed after removing the old tread by skiving on an inflated rim in order to achieve standard road wheel dimensions. The buffed casing should have dimensions compatible with the retreading system used. The worn tread surface is removed to obtain a symmetrical profile in accordance with procedural specifications. The buffed area of the casing should be free of contamination and oxidation. All buffing should be performed on an inflated buffer.

The worn tread, together with its under tread, is removed in a grinding operation using a suitable machine and specific grinding tool in order to prepare the old casing to have the required dimensions and surface consistency. The principal objective of buffing is to ready the tread area by removing worn-out tread for the application of new tread and to facilitate applying a new tread assembly. A buffing tool assembly made of high speed steel for an OTR tyre is shown in Figure 3.5.

It is essential to shape and size the tyre crown in order to prepare for the foundation of a new construction. A careful inspection is carried out after buffing to ensure that no unexposed overlooked defects remain. The dimensions of the tyre after buffing have to be appropriate for the type of tyre and size of curing mould, as well as the retreading system. To avoid any contamination, the time gap before assembly is kept to a minimum.

During buffing, the old casing must be buffed to a predetermined crown width, radius and profile proportion. Exposed frayed ends of fabric or steel cord must be removed and finished appropriately. On completion of buffing, the finished cords/steel should be coated with rubber cement and if any treatment is necessary to improve adhesion this should be carried out immediately after buffing. Any lapse of time between the final buffing operation and application of cement for the subsequent fabrication process may create adhesion problems.

Figure 3.5: Tyre grinding tool assembly made of high speed steel.

There may then be some chance of finding grooves, irregular wear spots and untexturised areas after removing the tread during skiving. Such surface blemishes are treated by a hand-operated grinding machine. Hand-grinding of such minor blemishes on a tyre is shown in Figure 3.6.

Figure 3.6: Hand grinding of blemishes on a tyre surface.

The prepared tread surface, where retreading is to be performed, should have an appropriate texture and consistent profile. A contamination-free buffed surface is essential for improved adhesion. A properly buffed tyre ready for the building-up process is shown in Figure 3.7. Every cut, hole and penetration must be checked to ascertain the severity of the damage and to ensure that all foreign material is removed.

Figure 3.7: Buffing operation completed for a 35 tonne dumper tyre.

On visual examination after buffing, if there are any damaged sections on the tyre surface that are bigger than 10 mm in size, they should be repaired using a suitable rubber compound before the building-up operations. After buffing, a thorough inspection is essential to ensure that there are no hidden and unnoticed defects on the tyre such as cuts, holes or injuries, or any foreign matter that should be removed. Photographs of pitholes and a damaged side wall surface exposed after buffing are shown in Figures 3.8 and 3.9.

3.4.2 Casing repair

All casing repairs must be conducted in accordance with prescribed methods and using suitable tools after buffing. Damage caused to a worn tyre during its life in service may be uncovered during buffing. Repairing damaged areas such as cuts, tears or dents is required, by means of moderate grinding using a motorised hand-grinding tool with controlled speed for thorough cleaning of the spot. The areas are

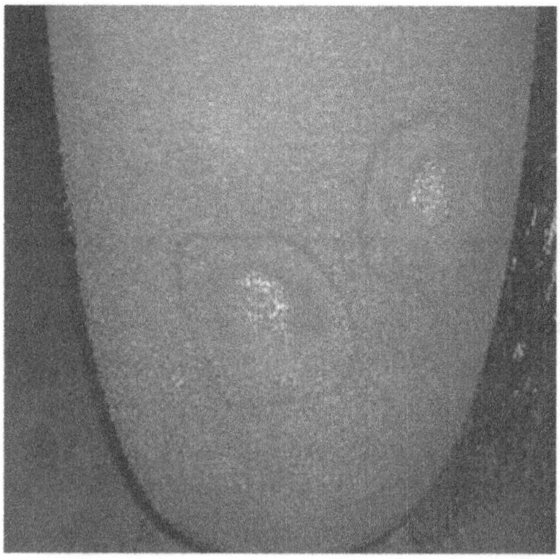

Figure 3.8: Dents needing repairing.

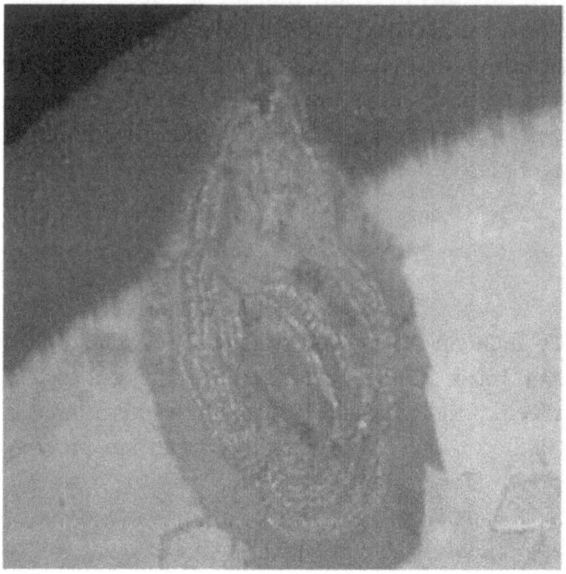

Figure 3.9: Damaged part after buffing needing repairing.

then applied with rubber cement. On drying of the cement, the areas are filled with prepared uncured repair rubber compound, so that the top tread can be applied without difficulty. Trained operators are needed to perform repairing jobs according to industry standards. The final determination of reparability, type of repair and repair material must be made after skiving and inspection, in accordance with the recommended specification and repairing criteria prescribed by the manufacturer of the repairing materials [9].

3.4.3 Cementing

Buffed casings are cleaned with an air jet to remove any dust from the surface and then cleaned with a suitable solvent. They are then allowed to dry and are cemented by applying an appropriate rubber-based cement of a specific viscosity within maximum of 8 h of buffing. The casing must be clean and free from foreign materials such as buffing dust, dirt and oil before application of the cement.

Buffed tyres ready for cementing should also be free from loose cords, and any exposed steel should be cemented within 15–20 min of its exposure after buffing. Rubber-based cement prepared in house should provide adequate adhesion strength between the new tread assembly and the prepared old casing. An acceptable drying time/tack-free time is determined in accordance with standard cement preparation before the assembly of other components.

Cemented tyres must be handled carefully to ensure that the surface is not contaminated with dust, which may impair adhesion property. They should therefore be stored in a dust-free, humidity-controlled room. It may be necessary to cover up the cemented tyre before the subsequent building up operation if the tyre is kept for a long period. The buffed tyre may need to be cleaned with a suitable solvent before building up, if it is stored for a longer period.

3.4.4 Building (Assembly)

Tyres ready for retreading must be built on an inflated rim to match road wheel dimensions. Extruded green rubber tread profile is centred on the buffed circumference of the worn buffed surface of the tyre, and tread pattern interruption is minimised with the tread splice. Both cut ends of the uncured tread are properly joined and adhere to the ends *via* filler rubber over the surfaces, and are free of contamination. The splice is cured together using suitable rubber-based cement. Tread stretch requirements and building tolerances should conform to industry accepted standards. Before the application of tread, gum or cushion gum sheeting is applied to the cemented casing; this is directly applied onto the treated surface from an extruder or the strip may be cut from calendered sheeting.

Tyres must be built on an inflated rim to match the wheel dimensions. The tread rubber should have a crown width, base width and gauge dimension suitable for the mould design and the size of tyre to be cured. The applied tread rubber should provide the required under tread gauge, and the rubber bonding surfaces should not be contaminated during handling. The extruded tread profile must be centred on the buffed circumference of the tyre. Stitching and splicing is carried out in such a way as to eliminate trapped air pockets without pulling the tread off centre or distorting the rubber. The tread width used is determined by the casing dimensions [10]. Images of extruded rubber profiles for retreading and the method of application are shown in Figures 3.10 and 3.11.

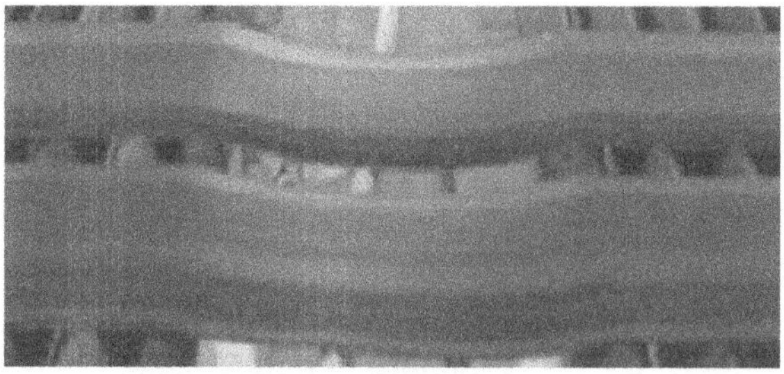

Figure 3.10: Extruded rubber retread profiles.

Figure 3.11: Application of extruded tread profile on buffed tyre casing.

3.4.5 Mould curing of retread (Hot process)

A fully assembled built-up tyre ready for hot curing in a moulding press should be stored to prevent distortion of the uncured materials, and should be kept free of any contamination. Curing procedures must follow specifications on curing time, temperature and pressure. Rubber curing tubes and rims appropriate for a requirement should be inserted properly while curing the assembled tyre in the circular press used for moulding. The total curing process is conducted in the moulding press, which is heated with steam to a temperature of 150–160 °C. The duration of curing varies from 2 to 8 h, depending on the size of the tyre. Air is injected under pressure into the inner rubber curing tube during vulcanisation; this is in order to exert a pressure from inwards to the mould walls. The air pressure should be maintained at 800–950 kPa throughout the curing process (depending on the size of the tyre retreaded).

The application of mould release agent (by spraying silicone emulsion into the inner periphery of the preheated mould) enables the quick release of the moulded tyre after curing, and reduces the likelihood of adhering to the mould surface. The outer periphery of the retread is designed to have the desired tread pattern according to the mould design.

The casings used for uncured tread rubber are uniformly buffed using a template or other system to ensure matching of the mould contour and size. The casing must be buffed properly to fit the mould in accordance with bead-to-bead measurements, diameter and section width requirements to conform to the specifications for the equipment used.

Mould release agent (a water-based silicone emulsion) is sprayed onto the preheated mould before placing the built-in tyre, and allowed to dry. This leaves an extremely thin coating on the mould surface and gives a better surface appearance on the clean and flawless lug designs of the finished retreaded tyre, and also makes it easy to release the cured tread from the mould.

The moulded retreaded tyre is allowed to cool at room temperature for several hours, and after cooling each tyre is tested with 200 kPa air pressure (using an inner rubber tube) before final inspection and dispatch [11]. Images of an assembled tyre moulded in a steam-heated press and a hot-moulded retread 35 tonne tyre are shown in Figures 3.12 and 3.13.

3.4.6 Precure retreading (Cold curing process)

The 'cold' curing method of tyre retreading is simpler than the 'hot' process involving building and curing. The tread surface of the casing is buffed properly after skiving. A fast-cure cushion gum film is applied on the surface of the tyre after application of rubber cement to achieve improved bonding of the new tread with the worn tyre. Precured tread that matches the tyre is placed on the outer circumference of the used prepared tyre. The assembled tyre is sometimes wrapped in a thin rubber envelope to accomplish better bonding with the old casing, although lighter vehicle tyres may be cured without the envelope and sent to an autoclave for vulcanisation, usually in batches.

Figure 3.12: Retreading moulding press.

Figure 3.13: Moulded retread for a 35 tonne dumper tyre.

This technique for the cold process of tyre retreading has proven its durability, and the tyre casing is less likely to degrade because vulcanisation of the retreading is performed at a lower temperature (in the range of 95–120 °C).

Following the cold curing system, the worn tyre casing is buffed to a width that is compatible with the casing dimensions. All tread grooves should be removed from the buffed area, and the buff should be centred on the casing with equal shoulder heights over the bead bundle area. The buffed tyre should be as close as possible to the specified radius. Radial tyres are buffed to a maximum of 3 mm remaining under tread over the belt package.

The following steps comprise a cold retreading system:
- Inspection
- Skiving and buffing
- Drying of casing
- Repair
- Cementing
- Building
- Enveloping
- Vulcanisation
- Final inspection

3.4.7 Precured retreading process

Appropriate precure tread is selected in strips or full circles with various designs, widths and thicknesses. It is then adhered to the casing using cushion gum, which acts as a bonding layer. It is essential to remove any air trapped between the tread and cushion gum layer during building, because this may create a non-contact area, which may cause ultimate failure of the retread top.

The correct tread width is applied to the prepared casing in a straight and even fashion. On application of the precured tread, the two cut ends of the tread are coated with vulcanisable rubber-based cement. After drying of the cement, a sizable cut strip of calendered cushion gum sheeting is applied between the joining ends, which are properly joined by a hand-operated stitching roller (metal roller). The assembled tyres, in batches, are then vulcanised in an autoclave. The process of laying precured tread on a make-ready tyre is depicted in Figure 3.14.

3.4.8 Enveloping

Following the cold retreading process, the next step after applying the precured tread on the tyre is to insert a rubber restraining device, which is designed to fit tightly around the tyre on the casing during autoclave vulcanisation. The envelope is fitted using a valve to creating a vacuum before vulcanisation, and removed by introducing air after curing. On evacuation of air it creates pressure on the built-in precured tread, enabling proper bonding of the new tread with the old casing. Sometimes, a liquid release agent is used on the outer surface of the envelope for ease of inserting and removing it. A rubber envelope used in cold retreading vulcanisation is presented in Figure 3.15.

Figure 3.14: Application of precure tread on a tyre.

Figure 3.15: Envelope used in cold retreading.

3.4.9 Vulcanisation of cold cure retreads

The curing chamber used for vulcanisation in a cold retread system is heated by pressurised steam/hot air using an electric heater or by hot oil passing through coils, and pressure is applied in the vulcaniser with pressurised air. Tyres, to be cured in batches, are stored in such a way as to prevent distortion of the uncured materials, and are kept free from contamination.

Envelopes, tubes or sealing ring devices used to cure a tyre should be free of leaks and defects. Multiple prepared and enveloped ready-to-cure tyres are suspended from a rail system in the autoclave and the door of the autoclave is closed. The curing procedures must conform to specifications on curing time, temperature and pressure. The autoclave operating pressure is usually maintained at a minimum of 0.3 MPa and the temperature in the range 100–120 °C, depending on the type of curative system used in the composition of the cushion gum. A vulcaniser for the cold cure retreading process is illustrated in Figure 3.16.

Figure 3.16: Vulcanising chamber for cold cure system.

After curing, the tyres are inspected and can then be used as a new retreaded tyre. The cold curing method is simpler than the hot curing system and requires less investment, but its material costs are higher.

3.4.10 Final inspection

The finished retread tyres finally undergo a hands-on inspection by a trained technician for quality assurance in relation to their service conditions, to conform to all specified standards followed by established retread industries. This includes a visual inspection for correct adhesion of the tread, and to asses curing and repaired spots. X-ray inspection is also carried out to detect any imperfections on the retread, sidewall and interior of the tyre, if such a facility is available to the manufacturer [12].

3.4.11 Exterior painting

After final inspection, the cured and finished retreaded tyre may be applied with a room temperature-cure rubber-based paint as a protective coating and to improve its aesthetic appearance for commercial reasons. Application of the paint may be by spraying or brushing. The finished tyre is then marked with an appropriate label indicating tyre size, date of retreading, guidelines for safe use, and so on [13, 14].

3.5 Advantages and disadvantages of cold *versus* hot retreading

This section provides a list of advantages and disadvantages of cold *versus* hot retreading [15].

Cold process:

- The cold retreading process entails vulcanisation without a mould at a temperature between 95 and 120 °C.
- It requires less investment, smaller production units, no expensive moulds and lower follow-up costs.
- The precure retreading requires high-quality carcasses.
- Material costs are higher than those for hot retreading materials.
- A wide range of tread types are available, allowing the optimum tread to be selected for the tyre application.
- There are low logistics costs.
- Repairing of the carcass is not permissible in the cold retreading process.
- The precure retreading process is not harsh to the carcass, because the vulcanisation temperatures are lower and put less strain on the rubber–metal bonds.
- Heat build-up, rolling resistance and other tyre properties are often easier to assess.

Hot process:

- Hot retreading requires vulcanisation of a tyre in a mould at a temperature of 150–165 °C.
- A separate mould is required for each tread and size. High investment is required to provide a range of moulds, and they need to be updated regularly.
- It is suitable for different tyres, sizes and applications, from car to aircraft tyres.
- Material costs are lower than the more complex products required for precure retreading.
- The production process needs to be designed for large numbers of tyres. This calls for a central production workshop, with relatively high logistics costs.
- Hot retreading allows extensive repairs, and can be carried out on the tyre carcass. Even bias-ply carcasses can undergo hot retreading without any problems.
- Hot retreading process may accelerate ageing of the carcass with the long duration of cure at higher temperatures.

3.6 Total tyre retreading logic

'Total tyre logic' was established to engineer earthmover tyres. It is a process of monitoring and managing the condition of tyres and also encompasses tyre supply, the survey of sites and suggests operational practices. The outcome is total management of the tyre and perceptible management of the asset.

In view of the fact that a tyre is a complex and high-tech safety product, subsequent testing of a retreaded tyre while in service, where the tread makes the only contact with the road, depends on varying factors that have an effect on the test results. These factors include the condition of the road surface, temperature of the terrain, speed of the vehicle, total load, acceleration, development of heat during running, alignment of the wheels, tyre air pressure and so on.

A tyre tread performs several functions while the tyre is running, including sustaining load, supporting low and high speeds, achieving and maintaining balance within a series of parameters without creating too high a rolling resistance (whether the ground is dry, wet, clean, asymmetrical or distorted).

This is a specially formulated tyre management curriculum/method exclusive to users deployed tyres, which acts to augment the life of a tyre during its service life. It is necessary to follow the approved specification, and regularly conduct preventive maintenance. Overall, it makes savings in terms of operating cost, repairs and the time for retreading.

Total tyre logic was developed as a programme to monitor and maintain OTR tyres such as dumper tyres, earthmovers tyres and so on. Useful recommendations are made for operational practices to result in effective tyre management, and efficient repairs and on time retreading. Total tyre logic imparts advice on operational performance, its appropriateness and monitoring of a tyre's wear and tear. It is suggested that when the tyre is worn to 2 mm it must be replaced with a newly retreaded tyre to prevent possible skidding and hydroplaning.

An online total tyre logic programme provides detailed information to the tyre user/ fleet owner about the tyres, any time and any place. The database collects information about the condition of the tyre and supplies this to the consumer.

Good quality OTR tyres and aircraft tyre casings are designed to last longer than a single tread life, and can be retreaded at least three or four times. The tread is the only consumable part of the tyre, and if the tyre is maintained, monitored, and removed and retreaded at the right time, a great saving can be made in operational expenditure.

This tyre management system also makes noticeable savings in terms of tyre casing scraps, thus reducing concern about the environment. Managing tyres prudently will have an improved effect on the control of environmental effluence.

Total tyre logic is a bespoke management programme targeted at a customer's particular tyres. It aims to prolong tyre life, initially by providing the correct specification, and then *via* preventive maintenance. Standards established for the testing of tread wear include measurement parameters for the tread loss profile, lug count and heel–toe wear. At present there is a global shortage of earthmover tyres, so the implementation of total tyre logic in the management of tyres, and thus retreading, will aid the crisis.

Effects of sporadic wear of the tread are accountable for irregularities and its management system:

- *Wheel alignment*: This leads to disproportionate wear of the inner and outermost ribs of the tyre, and irregular wear of the tread. A sudden bump from a pothole can cause misalignment and impairment of the tyre. Periodic checks of the alignment will ensure proper maintenance of the tyres.
- *Tread wear*: Anomalous wear of the tread can decrease its road gripping ability in unfavourable conditions. Periodic visual monitoring of the tyre condition for inconsistent wear of the tread will indicate irregular performance of the tyre in maintenance.
- *Rough and rocky haul road*: This sort of terrain will speed up excessive wear to the outer ribs. Steel-belted radial tyres help in preventing this category of tread wear.
- *Tyre pressure*: Under-inflated tyres lead to excessive stress on the tyres, irregular wear and loss of control, and increase the chance of an accident. Tread life will decrease in under-inflated tyres due to wear of the tread on the edges or shoulders. Under-inflated tyres generate excessive heat, which also lowers tyre life. Furthermore, by increasing the rolling resistance they reduce fuel economy.

In an over-inflated tyre, wear takes place at a faster rate at the centre than at the outer edges, thereby reducing the life of the tread. Regular tyre pressure monitoring at regular intervals is therefore a requirement in tyre management system. Figures 3.17 to 3.19 illustrate the wearing of tyres that have not been properly inflated at the specified pressure [16, 17].

3.7 Cause and effect analysis for tyre retreading process

In expressions, the cause and effect of an inference of affirmation describing the purported cause to assertion relating to the professed effect comply with specific conditions. The observations of cause and effect relates to placing different conditions acceptable as worthy inference. This provides a method of reasoning about cause and effect that is clear and useful for the industry. Consequently, cause and effect analysis of the tyre retread industry will

Figure 3.17: Correctly inflated tyre.

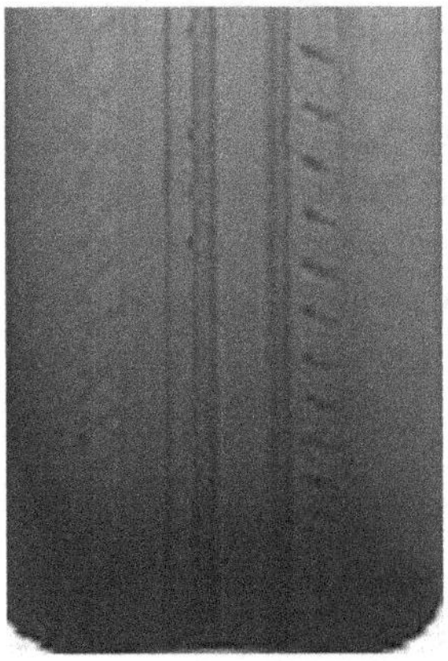

Figure 3.18: Under-inflated tyre, showing wear on outer edges.

highlight all the factors that are essential to managing trouble-free operation of a plant.

The quality of a retreaded tyre determines the physical attributes of the end product, so the production process requires a system that can produce good quality product. There are number of quality improvements methods and approaches that have been applied to improving standards in its process and products. Quality assurance curricula and monitoring systems have been introduced in many manufacturing industries, and the tyre retreading industries are no exception in implementing these, with assured results.

Tyre retreading manufacturing is effectively a re-engineering activity and the main focus must be to satisfy customers by offering quality products, timely delivery and after sales services at equitable cost.

Figure 3.19: Over-inflated tyre, showing wear in the centre.

The gnawing away of a retreaded tyre during its service, which may affect different aspects of application, processing, maintenance and the tyre itself, has been analysed, and the probable effects are shown in Figure 3.20 [18].

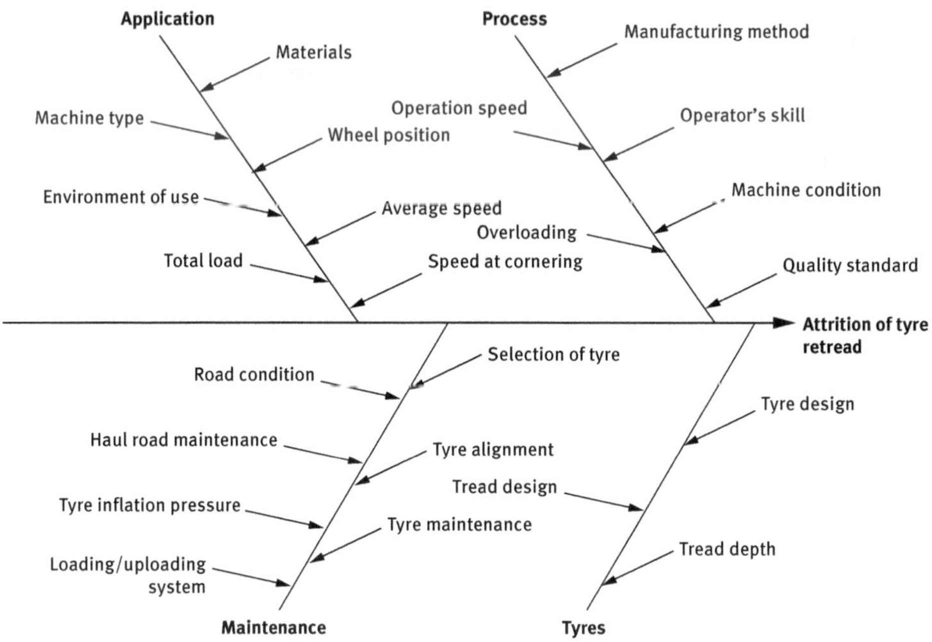

Figure 3.20: Dynamics of a tyre life – cause and effect.

References

1. B. Banerjee and S.K. Pal in *Retreading of Tyres – Methodology and Its Significance in Economics and Environment*, Book of Abstracts, India Rubber Expo – Conference and Workshop, Mumbai, India, 2013, p.189.
2. P.W. Dufton in *End-of-Life Tyres – Exploiting Their Value*, Rapra Technology Ltd, Shawbury, UK, 2001.
3. V.A. Gudkov, V.N. Tarnovskii and S.V. Ganzin, *International Polymer Science and Technology*, 2001, **28**, 9, 21.
4. D.L. Bender in *Tire Retreading the Rationale and a Review of Technologies*, Rubber World, September 2007, p.1.
5. *Science and Technology of Rubber*, 3rd Edition, Eds., J.E. Mark, B. Erman and F.R. Eirich, Elsevier Academic Press, London, UK, 2005.
6. M.S. Evans in *Tyre Compounding for Improved Performance*, Rapra Review Report No.140, Rapra Technology Ltd, Shawbury, UK, 2001.
7. *Rubber Basics*, Ed., R.B. Simpson, Rapra Technology Ltd, Shawbury, UK, 2002.
8. *Handbook of Rubber Bonding*, Ed., B. Crowther, Rapra Technology Ltd, Shawbury, UK, 2001.
9. *Retreaded Tyres for Passenger Cars*, GB 14646–XXXX, National Standards of the People's Republic of China. http://www.puntofocal.gov.ar/notific_otros_miembros/chn142_t.pdf.
10. eHow http://www.ehow.com/about_5109896_tire-retreading.html.
11. Retread Manufacturers Association. http://www.retreaders.org.uk/.
12. Tire Retread & Repair. http://www.retread.org/.
13. *Recommended Specification Guidelines*, Section 11: Retreading and Section 13: Standards & Regulations, Goodyear. http://www.goodyeartrucktires.com.
14. TireBusiness.com.http://www.tirebusiness.com/.
15. *Retreading Process*, European Tyre & Rubber Manufacturers' Association (ETRMA), Brussels, Belgium, 2006.
16. OTR Tyre Retreading, *Mining Weekly*, 12th May 2006. http://www.mining-technology.com/.
17. Kal Tire Mining Tire Group. http://www.kaltiremining.com/en/.
18. *Safety and Maximization of Tire Life*, Agora Tire Inc., Kingwood, TX, USA, 2008. http://www.agoratire.com/.

Tuhin Chatterjee and Kinsuk Naskar

4 Green tyre tread technology

4.1 Introduction and historical overview

Historically, the spelling was 'tire', and it came from the French word 'tirer', meaning 'to pull'. Indeed, 'tire' originally referred to the iron hoops or thick wires bound to carriage wheels. In French blacksmithing the word for a drawn iron rod is a 'tirer', or 'pull'.

In 1845, Robert William Thomson, a Scottish engineer, first invented the pneumatic tyre. More than 40 years later, in 1888, a pneumatic bicycle tyre made from natural rubber (NR) was introduced by John Boyd Dunlop to make his son's bicycle journey more comfortable, leading to the invention of the first commercially available pneumatic tyres [1, 2].

In 1915 Palmer Tyre Company of Detroit pioneered the first rubberised 'cord' fabric and made the first 'cord tyre'. By 1937, steel cords were being used in the manufacture of cross-ply truck tyres. It should be remembered that right up to the beginning of World War II, many trucks still used solid tyres.

In 1948 Michelin developed the first radial tyre. This was revolutionary, as they were steel-belted radial tyres. They had the advantages of longer life and increased mileage. However, they required a different suspension system and were therefore only adopted gradually. This tyre, along with Dunlop's invention, gave us the tyre we have today [2].

In 1904, the introduction of carbon black as a reinforcing agent for rubber led to improvement in the wear resistance property of tyre tread [3–5]. Generally, carbon black is used for reinforcement purposes and enhances the physical properties of the filled rubber compound. The high loading of carbon black also reduces the amount of zinc oxide needed. This was previously used to enhance the rubber's physical properties, although a small amount of zinc oxide is still added with the sulfur for accelerator-activated curing purposes. Conventionally, pneumatic tyre tread has been prepared by blending various rubbers and is typically reinforced with carbon black.

In the late 20th century, further important progress in tyre technology was made with the replacement of carbon black by a group of highly active silicas. In 1992 Michelin introduced the concept of 'green tyre technology', using silica filler and a bifunctional coupling agent rather than conventional carbon black. In tyre manufacturing, silica is used mainly to reduce rolling resistance. The increasing importance of protecting the environment has given rise to a demand for tyre tread that combines a long service life with driving safety and low fuel consumption, and this is achieved by lowering the rolling resistance. The use of silica therefore leads to lower fuel consumption and better mileage than tyres with treads made from compounds containing carbon black. Technological reasons have long prevented

https://doi.org/10.1515/9783110643893-004

silicas from being used in tyre tread compounds. An insufficient degree of physical or chemical bonding between the silica particles and rubber matrix has prevented silica fillers from being used as reinforcing fillers for rubber for the most purposes, including tyre treads. To overcome this, coupling agents that react with both the silica surface and the rubber matrix are often used [6].

Today, silica-treated carbon black (carbon–silica dual-phase filler, CSDPF) has also been investigated as a novel reinforcing filler for tread compounds. CSDPF functions just like silica in tread compounds [7].

The tyre tread ensures grip between the road surface and the tyre. After rolling 1,000 times the tread depth reduces, as does the coefficient of friction, leading to slipping between the road surface and tread. At this point the tyre is no longer in a condition to be used. In the last 20–25 years, the retreading process, a modern approach, has gained increasing importance due to its cost-effectiveness, facility, and greater profit margin than the full replacement of a tyre.

4.2 Reinforcement of rubber

Elastomers or rubbers (NR or synthetic) are not usually used in their pure form due to a lack of hardness, strength and wear resistance. Generally, elastomers are reinforced with fillers to meet practical requirements. The addition of fillers fundamentally changes the properties of the rubbers. According to the kinetic theory of elasticity, the rubber modulus (with no fillers) increases with the rise of temperature. The addition of fillers significantly changes the temperature coefficient of modulus; it can even alter the sign of the coefficient, resulting in a decrease of the modulus with increasing temperature. Fillers such as carbon black and silica are generally used to improve the properties of rubber compounds. Reinforcing fillers gives the material unique properties: a combination of high elasticity with high strength. In retreading, the addition of an increasing amount of filler also reinforces the retread compound and in some cases improves adhesion of the retread compound with the old tyre compound. The effects of increasing the amounts of reinforcing fillers on different properties of the elastomers are shown in Figure 4.1.

4.3 Reinforcing fillers

The rubber compounds used for tyre retreading are more than just an elastomer that can be formed into a fixed shape by crosslinking, but contain many different compounding ingredients, such as processing aids, antioxidants, antiozonants, curing agents, accelerators and a considerable amount of filler. Fillers are used for

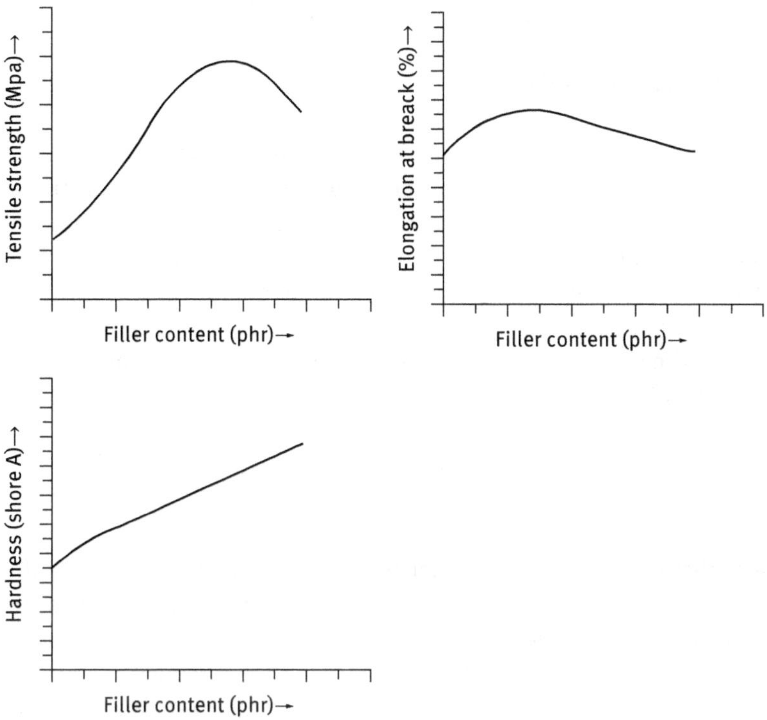

Figure 4.1: Influence of reinforcing fillers on the properties of an elastomer.

many reasons, including (i) reinforcement, (ii) improving processing, and (iii) as a cheapener. The classification of filler is made on the basis of chemical composition and influence on rubber properties. Fillers used in rubber compounding canbe divided primarily into three categories with respect to their reinforcing effect:

- Inactive
- Semi-active
- Very active

The term 'active' or 'reinforcing' refers to the influence of the filler on compound viscosity and mechanical properties such as tensile strength, abrasion and tear resistance. The main characteristics that determine the reinforcing effect of the filler are its structure and surface properties. 'Active' fillers have a large relative surface area and high structure, giving a strong physical and chemical interaction between the filler and polymer. However, a 'highly active filler' surface also results in strong interparticle forces, which negatively influence processing behaviour due to the agglomeration of filler particles during mixing and storage [8].

4.4 Factors influencing filler reinforcement

4.4.1 Particle surface activity

Active fillers can provide a stronger interaction between the filler and rubber matrix than inactive fillers. The chemical interaction between the filler and rubber matrix significantly depends on the type or form of the functionality on the filler surface [9]. Sometimes, modification of the filler surface is needed in order to match the surface chemistry to a specific rubber. Surface modification of a filler improves its adherence with rubber. In retread compounds, a modified filler surface sometimes improves the adhesion of the retread compound with the old tyre. Silica is one such active filler that requires surface modification.

4.4.2 Particle size

The key parameter that strongly affects rubber properties is the particle size of the filler particle. Carbon black and silica have very small primary particle sizes in the range of 10–100 nm and are classified as reinforcing fillers, which can significantly improve elastomer performance. Semi-reinforcing and non-reinforcing (diluent) fillers have a particle size in the range of 100–1,000 mm and 1,000–10,000 mm, respectively. In fillers with particle size greater than 10,000 nm, this greatly exceeds the polymer interchain distance, resulting in an area of localised stress, leading to rupture of the elastomeric chains on flexing or stretching and hence deterioration of the rubber properties. The classification of fillers based on their particle size is presented in Figure 4.2 [10].

4.4.3 Particle surface area

The reinforcing fillers used in retread compounds have a high surface area, providing a greater number of active contact sites and giving a strong filler–rubber interaction. There are three methods that can be used to measure the particle surface area of fillers:

1. Nitrogen surface area (NSA)
2. Statistical thickness surface area (STSA)
3. Cetyl-trimethyl-ammoniumbromide (CTAB) adsorption method

Methods (1) and (2) are based on the Brunauer–Emmett–Teller (BET) nitrogen-adsorption theory [American Society for Testing and Materials (ASTM) D6556]. Nowadays, measurement of carbon black surface area using the CTAB method has been replaced by the STSA technique, because the CTAB method is labour-intensive and has poor testing precision [10], but the CTAB method is still applicable for the measurement of

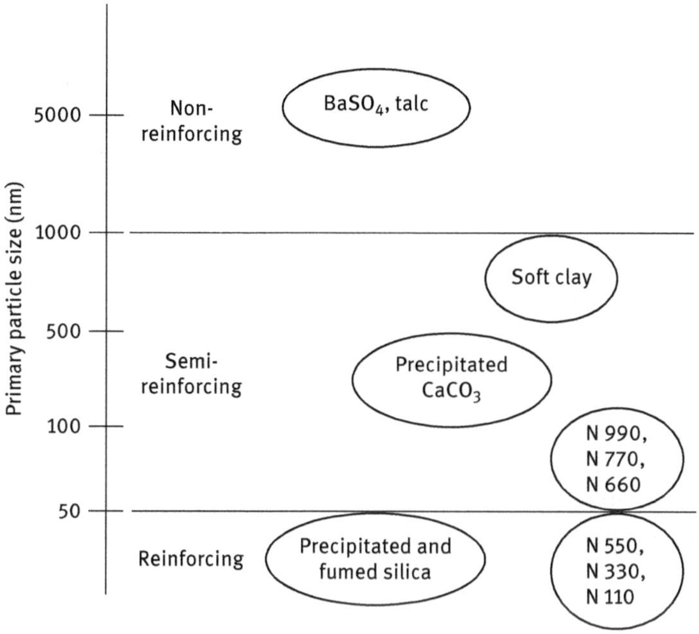

Figure 4.2: Classification of fillers based on the particle size of primary filler particles.

the surface area of silica fillers. The NSA (or BET) method measures the overall surface area of the filler particles, including micropores (pore diameters <2 nm). The STSA and CTAB adsorption methods, respectively, determine only the external surface area of the carbon black and silica fillers, and the values obtained are assumed to be the specific surface areas of filler particles that are accessible to the rubber molecules [11].

4.4.4 Particle structure

The primary particles of reinforcing fillers (such as carbon black and silica) are spherical. However, during production, aggregate or agglomerate formation takes place due to self-association of the active functionalities on their surfaces [12]. The anisometry of these fillers is described in terms of 'structure', which includes aggregate shape, density and size. During the compounding of retread compounds, a higher structure of filler is better from the point of view of reinforcement efficiency [12, 13].

4.5 Reinforcing effects

In retread compounds, filler reinforcement originates from the interaction between the filler particles and rubber. These interactions may be strong (e.g., covalent bonds between functional groups on the filler surface and the rubber) or weak (e.g., physical attractive forces). The level of physical interaction is high when carbon black is blended with a rubber, but the interaction between silica particles and the rubber is very weak and the bond is formed between the filler and the rubber only with the use of a coupling agent. The reinforcement of rubber is often defined as the improvement in properties such as an increase in modulus, tensile strength, wear resistance, tear resistance, and so on. In practice, the definition of reinforcement is the improvement in the service life of rubber articles (which could fail in different ways).

Above a critical concentration threshold (percolation threshold) of filler particles, the interaction between the filler particles increases rather than the polymer–filler interaction, resulting in the formation of a filler–filler network. This filler–filler structure drastically changes the material properties and leads to an increase in the conductivity results of a carbon black-filled compound. In 1965, Payne first studied the strain dependency of filler-reinforced rubber to understand the reinforcement mechanism of filled rubber.

The main contributions to the complex shear modulus are (i) a strain-independent hydrodynamic effect of the filler; (ii) a strain-independent contribution of the rubber network; (iii) a strain-independent effect due to chemical/physical rubber-to-filler interactions; and (iv) a strain-dependent contribution of the filler, also known as the Payne effect (Figure 4.3) [14].

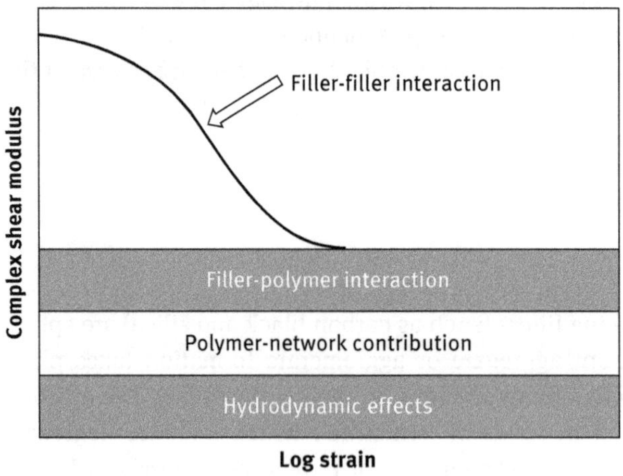

Figure 4.3: Parameters contributing to the complex shear modulus.

4.5.1 Hydrodynamic effect

This parameter is another strain-independent contribution to the modulus. This effect is described as an increment in the viscosity of the viscous fluids due to the addition of filler particles. In a polymeric matrix this effect is also measured as an increment in the moduli of the rubber compounds. The phenomenon was proposed by Einstein almost a century ago. He described the viscosity increase introduced by rigid spherical particles dispersed in a liquid by Equation 4.1 [15]:

$$\eta_f = \eta_0(1 + 2.5\phi) \tag{4.1}$$

This means that when a filler of volume fraction ϕ is dispersed in a fluid, the viscosity of the liquid suspension becomes η_f, where η_f is the viscosity of the fluid containing the particles, η_0 is the viscosity of the pure fluid, and ϕ is the volume fraction of the particles. For this model, Einstein made the following assumptions:
- Perfect wettability of the spheres
- Uniform spherical particles
- No interaction between the particles

The last two requirements, in particular, do not match actual filled-rubber systems, so, based on practical experience and Einstein's theory, Guth and Gold introduced an additional term and modified Equation 4.1 to give Equation 4.2 [16]:

$$\eta_f = \eta_0(1 + 2.5\phi + 14.1\phi^2) \tag{4.2}$$

The term Guth and Gold added is again a function of the filler volume fraction ϕ to the power of two, emphasising the influence of filler concentration.

Smallwood showed that, in the case of an elastic matrix, the viscosities of the material can be replaced by the shear moduli, as shown in Equation 4.3 [17]:

$$G_f = G_0(1 + 2.5\phi + 14.1\phi^2) \tag{4.3}$$

The addition of fillers increases the shear modulus of the pure elastomer G_0, which results in a shear modulus G_f for the filled compound. The same correlation holds true for Young's modulus E. A correction for the non-spherical shape of the filler particles, shape factor f_s, was added by Guth, resulting in Equation 4.4 [18]:

$$G_f = G_0(1 + 0.67f_s.\phi + 1.62f_s^2 .\phi^2) \tag{4.4}$$

Shape factor f_s represents the ratio of the longest dimension to the shortest dimension of the particle. The modulus, as calculated by Equations 4.3 and 4.4 is independent of applied strain.

4.5.2 Filler–polymer interaction

It has been shown experimentally that the storage modulus increases drastically with increased filler loading as a result of the filler–polymer interaction. Several bound rubber models have been proposed by researchers for filler–rubber interactions in carbon black-filled compounds.

Medalia and Kraus proposed the 'occluded rubber model' [19]. In this model, rubber chains are partly trapped in the voids of the filler aggregates in dependence on the aggregate and agglomerate geometry of the filler (Figure 4.4). In this way, rubber chains become immobilised and shielded from deformation. The rubber layer attached on the surface of the filler particles exhibits a glassy-state character. The immobilisation of the rubber is attributed to physical interactions such as van der Waals forces, or to chemical reactions (as in the case of the silica-coupling agent system), which give an 'in-rubber structure' (in other words a strong filler–rubber interaction). This occluded rubber does not contribute to the elastic behaviour of the rubber matrix, but acts as a rigid filler. Occluded rubber therefore increases the effective filler loading and thus the strain-independent contribution to the modulus.

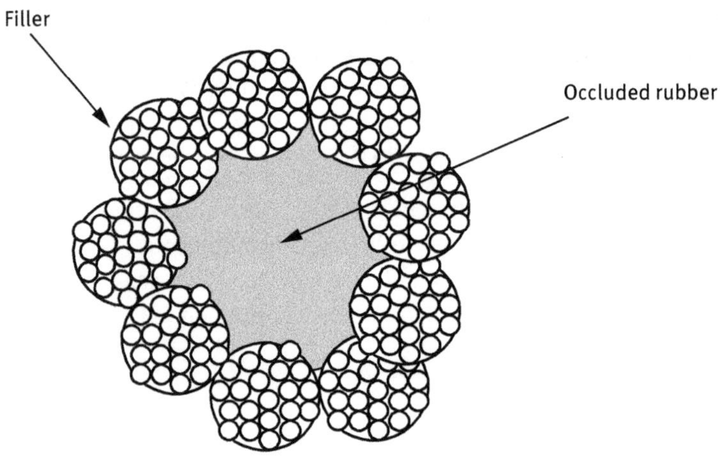

Figure 4.4: Occluded rubber model.

Smith and Pliskin proposed the 'shell rubber model' [20]. According to this a rubber layer can exist on the filler surface, similar to the chemical adsorption of a polymer on a filler surface, as shown in Figure 4.5.

Based on the shell rubber model, O'Brien and co-workers have developed the 'advanced bound rubber model' [21]. In this model, a double-layer structure of polymer composed of a glassy-state layer and a rubber-state layer is formed on the carbon black particles (Figure 4.6).

4.5.3 Polymer network contribution

Prior to vulcanisation, a small amount of polymer networking take place *via* physical crosslinks of entangled chains, and some chemical crosslinks occur during

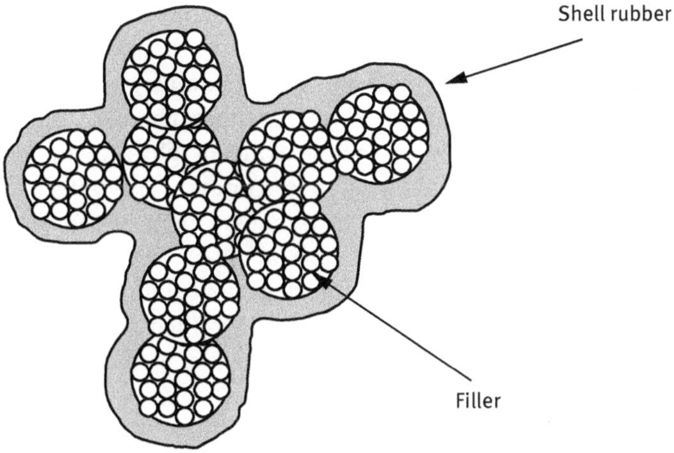

Figure 4.5: Shell rubber model.

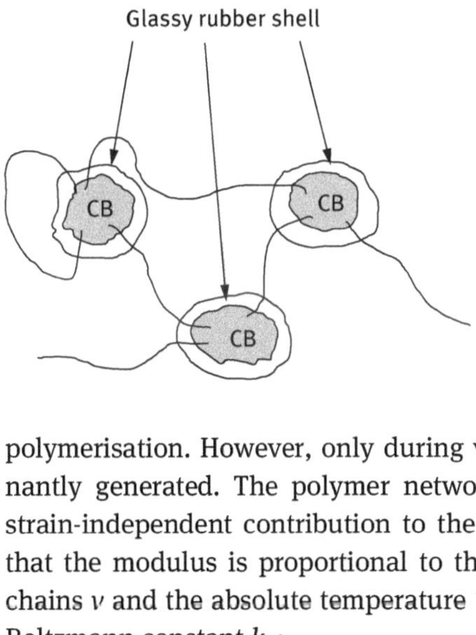

Figure 4.6: O'Brien model of bound rubber. CB: carbon black.

polymerisation. However, only during vulcanisation the rubber network is predominantly generated. The polymer network formed during vulcanisation is the third strain-independent contribution to the modulus. From Equation 4.5 it can be seen that the modulus is proportional to the concentration of elastically active network chains v and the absolute temperature T, with the proportionality constant being the Boltzmann constant k_B:

$$G_0 = v \times k_B \times T \tag{4.5}$$

4.5.4 Filler–filler interaction: Payne effect

According to Payne's interpretation, the filler–filler interaction in reinforced rubber causes a strain-dependent contribution to the modulus, the so-called

Payne effect. Payne demonstrated that there is a sigmoidal decrease in the storage modulus of highly filled rubber as the strain increases from a limiting zero-amplitude value to a high amplitude, and the loss modulus has a local maximum value at the strain where the storage modulus changes most. According to Payne this behaviour is due to the breaking of physical bonds between filler particles (e.g., van der Waals or London forces). This effect is strongly reversible once the strain is released, and is independent of the type of rubber, but is significantly dependent on the type of filler. The strain dependency of the Payne effect and the strain-independent contributions to the shear modulus for carbon black-filled compounds and silica-filled compounds are illustrated in Figure 4.7 [22].

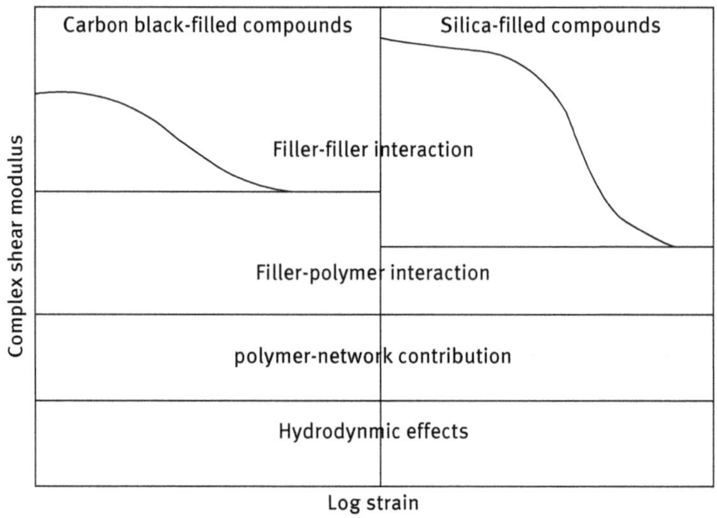

Figure 4.7: Comparison of shear modulus as a function of strain for carbon black- and silica-reinforced rubber vulcanisate.

4.6 Black and non-black reinforcing fillers

Reinforcing fillers for retreading purpose can be divided into two categories: (i) black and (ii) non-black. In general, carbon black falls into the first category, and silica, whiting, talc and so on fall into the second category [23].

4.6.1 Carbon black

Carbon black is mainly composed of carbon, with a low percentage of other hydrocarbons or oxygen-containing groups (e.g., phenol, carboxyl, ketone, lactone, pyrole or quinine). Carbon black is the most commonly used reinforcing filler for retreading compounds due to its high reinforcement capability. The surface is relatively non-polar, resulting in good compatibility with the hydrocarbon rubber. Although the chemically active sites comprise less than 5% of the total surface, reinforcing activity is achieved, mainly due to (i) mechanical interlocking of rubber onto the carbon black surface, (ii) chemisorption of rubber onto the filler surface, and (iii) van der Waals forces between the carbon black and rubber. Properties that improve after application of carbon black include (i) strength, (ii) abrasion or wear resistance, (iii) fatigue resistance, and (iv) ease of processing. However, because of its black colour it absorbs most ultraviolet components of sunlight, which can initiate oxidative degradation of the compound [5, 23].

4.6.2 Silica

Only black rubber articles can be prepared with carbon black. Accordingly, to produce highly durable coloured products, alternative active fillers were required, leading to the development of non-black fillers. In general, the surfaces of silica particles are covered by polar silanol (–Si-OH) groups, which are chemically active. The difference in polarity between the silica surface and the rubber hydrocarbons leads to low compatibility of silica fillers with general purpose rubbers and hence weak filler–rubber interactions [24].

4.6.3 Carbon black versus silica

Compared to carbon black, the Payne effect is stronger for silica, resulting in a weaker filler–polymer interaction and strong filler–filler interaction. As a consequence of this strong interparticle force between the filler particles, silica-filled tread compounds are characterised by (i) higher compound viscosity, (ii) lower bound-rubber content, and (iii) a higher modulus at low strain amplitudes and lower modulus at high strain amplitudes. However, the combination of a coupling agent with silica in retread compounds enables a better reinforcing effect and different dynamic mechanical properties than carbon black-filled compounds.

Silica has become more important in the rubber industry since the introduction of the Energy® tyre by Michelin. The main reason for this is the better reinforcing effect of silica-filled tread compounds in the presence of a coupling agent than with carbon black, and also the lower tan δ (loss tangent) value for silica at lower temperatures. The stronger reinforcing effect of silica allows a reduction of the filler content without any deterioration of the retread compound. On the other hand, a reduction in filler content adds a positive effect for elasticity because of the higher ratio of the elastic component of damping filler.

Replacing carbon black with silica filler therefore results in a lower tan δ value at a higher temperature and an additional reduction in the rolling resistance of the retread compound. A comparable tan δ at lower temperatures after the addition of silica provides a comparable ice grip and wet grip for the tyre.

In the case of carbon black, the formation of a co-continuous network leads to reinforcement. Upon repeated application of strain, the polymer network stores energy; disruption of the carbon black network takes place, leading to the release of heat energy. In retread compounds, compared to the case with carbon black, disruption of the silica-filled network generally takes place at higher strains. As a result, energy loss for a rolling tyre is less for a silica-reinforced compound than for a carbon black-reinforced compound, resulting in a lower tan δ and consequently a lower rolling resistance of the tyre [25].

It has been well established that the abrasion resistance or wear resistance properties of a tread compound are predominantly determined by filler characteristics such as surface activity and morphology. The high surface area of carbon black provides a better interfacial interaction with the polymer, resulting in a better wear resistance property. For all silica-filled compounds the abrasion index is 20–40% lower than that of carbon black compounds. However, compared to carbon black (with a similar surface area and structure), silica fillers have a poor polymer–filler interaction and therefore give poor abrasion resistance (Table 4.1) [26].

Table 4.1: Comparison of carbon black and silica.

Property	Carbon black	Silica
Abrasion resistance	Good	Bad
Wet skid resistance	Comparable	Better
Rolling resistance	Bad	Good

4.6.4 Various classes of silicas

Depending upon the manner of production and the resultant properties, two classes of silica are generally used in the polymer industry:

- Fumed silica or pyrogenic silica
- Precipitated silica

'Fumed silicas' are generally produced by the flame hydrolysis of silicon tetrachloride, as shown in Equation 4.6:

$$SiCl_4 + 2H_2O \rightarrow SiO_2 + 4HCl \tag{4.6}$$

Within the rubber industry, fumed silicas are nearly exclusively used in silicone rubber compounds, and are therefore of little interest in the present context.

Equation 4.7 shows that, in the presence of an acid (including mineral acids such as sulfuric acid, hydrochloric acid or carbonic acid), 'precipitated silicas' are prepared by precipitation from water-soluble sodium silicate ('water-glass'):

$$Na_2OSiO_2, XH_2O + H_2SO_4 \rightarrow SiO_2 \downarrow + Na_2SO_4 \tag{4.7}$$

A primary silica particle has cross-sectional dimensions of 5–100 nm, but aggregates formed by chemical and physical interactions of primary particles have dimensions of 100–500 nm. Aggregates are quantified according to (i) the specific surface area of primary particles, (ii) their geometrical arrangement, and (iii) the number of primary particles. The term 'structure' encompasses all three parameters and gives a general measure of the aggregate. Aggregates condense into agglomerates with typical dimensions of 1–40 µm. Agglomerates of silica more or less disintegrate to the size of aggregates or even primary particles during mixing with rubber. Silica aggregates are similar to those of carbon blacks, but their higher structure gives a greater reinforcing effect than carbon black. The higher specific component of surface energy of silica filler results in a stronger tendency to form agglomerates, difficulty in dispersing in the rubber, and even re-agglomeration after mixing.

During the development of highly dispersible silicas, as commonly used in modern low-rolling-resistance tyre tread, the best combination of aggregate and agglomerate size should be determined. In general, there should be a sufficient number of pores on the surface of the aggregates that the rubber can penetrate easily; this also increases the contact area between the silica and rubber. Stepwise precipitations at different pH values results in bimodal size distributions with different size ratios, and this provides lower rolling resistance to tyre tread. Bimodality is also generated during mixing by blending two fillers with different particle size distributions with the rubber compound.

Results have shown that a compound containing common commercial silica had approximately 6% area in a field of view with agglomerate size greater than 1 µm. This silica was therefore characterised as 'highly dispersible' silica, with a higher tan δ at 0 °C along with a lower tan δ at 60 °C Moreover, a low-viscosity retread compound is also obtained with a high dispersion of precipitated silica, which enhances mixing and extrusion processability. Consequently, low tread wear and high wet traction, together with low rolling resistance, were obtained.

Hi-Sil^{TM} EZ, an easily dispersible precipitated silica commercialised for use as a reinforcing filler of general purpose elastomers, affords excellent processability while maintaining important durability properties such as tear strength and cut-growth resistance [27, 28].

4.7 Silica reinforcement

Silica fillers are amorphous in nature. Their particle size is very small and they are generally classified into semi-reinforcing or reinforcing categories, like carbon black. There is enormous diversity in the types of silica, because of the different production processes and process variations. So, depending on the conditions during production, silica fillers can be of various types, as shown in Table 4.2 [29].

Table 4.2: Various classes of silica depending upon precipitation conditions.

Category	Required pH	Drying time	Dispersibility
Conventional silica	High	Long	Bad
Semi-HD silica	Low	Long	Moderate
HD silica	High	Short	Good

HD: Highly dispersible

Several attempts have been made to produce a classification of silicas (as for carbon blacks) by the International Organization for Standardization and the ASTM, but, due to the complicated nature of the silicas, these have never come into effect. In terms of the reinforcing properties of silicas, their particle dimensions, surface morphology, surface activity and bonding ability for coupling agents are of particular importance.

4.7.1 Aggregate size of silica

The microstructure of silica filler plays an important role in determining the physical properties of rubber retread compounds. Interaction between the primary particles may occur because of the close proximity of the particles. Primary silica particles form aggregates with typical dimensions of 100–200 nm, which are considered the real reinforcing species in rubber compounds. 'Structure' measurement can be done based on the adsorption of dibutylphthalate (DBP), so-called DBP adsorption. Conventional silica has a typical DBP adsorption value of 175/100 g (and 200/100 g or above for highly dispersible silica).

Another method is the so-called crushed DBP measurement, which is carried out while increasing the pressure on the silica from 0 to 30 MPa. This demonstrates that the structure of highly dispersible silicas is higher and they are less fragile than conventional silicas. Improvement in the dispersion characteristics of highly dispersible silicas may be explained by their high aggregate porosity, as they survive for longer than conventional silicas. The polymer has more space and time to penetrate into the voids in highly dispersible silica than in conventional silicas, which have a more compact structure. Furthermore, aggregates of highly dispersible silicas have a more branched structure, with three to four major branches on average [30], which also improves the dispersion characteristics of the highly dispersible silica during the mixing process and also reflects a bimodal distribution of the aggregates. Compared to conventional silica, the amount of small aggregates is relatively high in highly dispersible silica.

4.7.2 Specific surface area of silica

Two methods are generally used to determine the specific surface area of silica filler:
- N_2-adsorption, the so-called BET method
- *N*-cetyl-*N,N,N'*-trimethylammonium-bromide, the so-called CTAB method

Primary particle sizes of highly dispersible precipitated amorphous silicas are in the range of 10–50 nm, as determined by N_2-adsorption using the BET method. The BET method calculates the overall surface area of the particles, including the micropores. Commonly, micropores are small in size (<2 nm). No polymer molecules can penetrate into them. Indeed, only low-molecular-weight chemical compounds like vulcanising agents and coupling agents may penetrate, resulting in a loss of active components. Generally the BET surface varies between 50 and 300 m^2/g, and more commonly between 50 and 200 m^2/g. Due to their larger size, neither CTAB nor mercury can penetrate into the pores, so measure the external surface area of the ultimate primary particles. CTAB surface levels typically range between 100 and 200 m^2/g.

This means that CTAB values provide a more accurate indication of primary particle size than the BET surface area, and also strongly correlate with the physical properties of the filled rubbers, as depicted in Figure 4.8 [31].

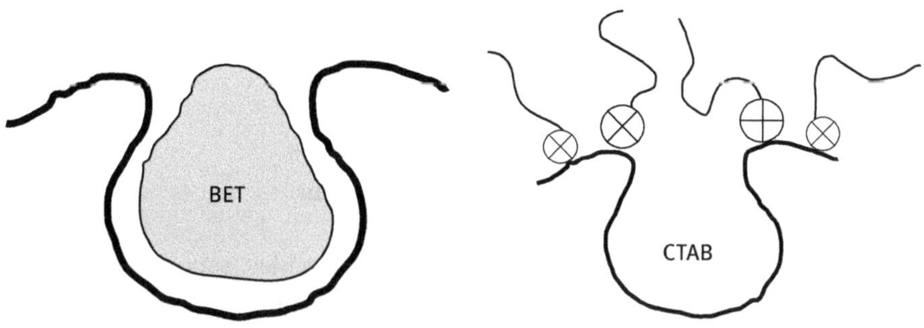

(a) (b)

Figure 4.8: Characterisation of filler surface (a) BET and (b) CTAB.

4.7.3 Characterisation of silica surface

Silica particles are generated from the polymerisation of silicic acid, forming complex amorphous polycyclic ring structures. Due to internal defects in these ring structures, silanol groups (≡Si-OH) and siloxane groups are generated on the surface of the silica particles. The silica surface is mainly characterised by (i) the number of silanol groups, (ii) the degree of hydration, (iii) the amount of adsorbed water and (iv) its surface acidity. These silanol groups are classified into three categories according to Si-nuclear magnetic resonance experiments or infrared spectroscopy, as shown in Figure 4.9 [30, 32].

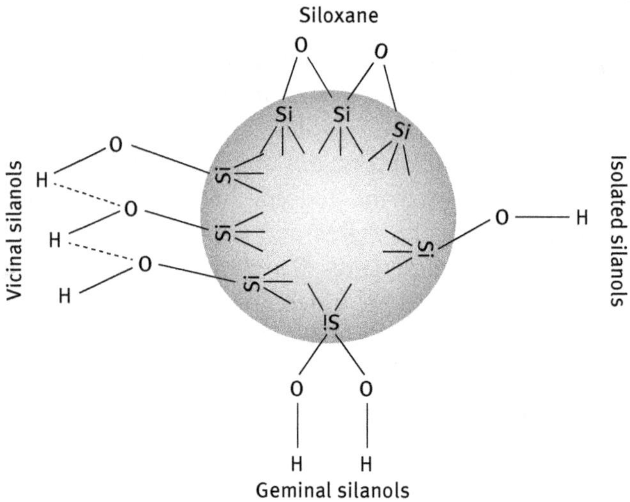

Figure 4.9: Types of hydroxyl groups on the silica surface.

Three different types of silanol groups are formed depending on the precipitation conditions:
- Isolated: a single hydroxyl group on a silicon atom
- Geminal: two hydroxyl groups on the same silicon atom
- Vicinal: two hydroxyl groups on adjacent silicon atoms

These silanol groups have a strong affinity for water molecules, especially adjacent silanol groups such as the geminal type of silanol groups, which are highly capable of absorbing water. In the case of highly dispersible silica, the germinal content is less than 20%. Silanol groups on the silica surface also improve the bondability of the retread compound with the old tyre.

4.7.4 Surface chemistry of silica

The previously mentioned, polymer–filler interaction strongly influences the reinforcing property of a compounded rubber. The surfaces of silica particles contain a large number silanol and siloxane groups, and can be characterised by the surface energy. Generally, the surface energy depends on the following factors: (i) dipole–dipole interactions, (ii) van der Waals forces, (iii) hydrogen bonding and (iv) electrostatic interactions [32]. In principle, surface energy is composed of dispersive and specific components and can be expressed by Equation 4.8 [33]:

$$\gamma_s = \gamma_s^d + \gamma_s^{sp}\,[\mathrm{J/m^2}] \tag{4.8}$$

where y_s^d is the dispersive component, which indicates the tendency to adhere to the polymer and y_s^p is the polar component, indicating the tendency of filler particles to interact.

For carbon black, the polar component y_s^p of silica is relatively high due to the presence of a large number of polar groups on the silica surface. The difference in the solubility parameters of the polymers and filler are responsible for the degree of wetting of fillers by polymers. The solubility parameters of some polymers and silica are given in Table 4.3 [33, 34].

Table 4.3: Solubility parameters of rubber and compounding ingredients.

Material	Solubility parameter
NR, BR, IIR	8–8.5
SBR	8.5–9
NBR	9.5–10
Carbon black	12–15
Clay	13–14
Silica (untreated)	14–18

BR: Butadiene rubber
IIR: Isobutylene–isoprene rubber
NBR: Nitrile rubber
SBR: Styrene-butadiene rubber

From Table 4.3 it is evident that, compared to polymers, the high solubility parameter of silica leads to difficulty in blending silica fillers with polymers. The compatibility aspects of silica fillers and model compounds that resemble the molecular structures of polymers can be investigated by an inverse gas chromatography study [35], and the compatibility of polymers and silica can be classified as follows:

High compatibility with silica: NBR > SBR > NR > BR > high vinyl BR > EPDM > IIR where EPDM is ethylene propylene diene monomer.

4.7.5 Bound rubber model of silica-filled rubber

During processing of silica-filled rubber, various chemical reactions take place and, as a result, a new advanced bound rubber model has been developed. Generally, there are two components: (i) occluded rubber in silica aggregates and (ii) cross-linked polymer due to polymer chain scission and recoupling – that form the bound rubber of silica-filled rubber [36]. Luginsland and co-workers [35] proposed a simple model of silica/silane reinforcement based on the hydrodynamic–occlusion–interaction theory proposed by Medalia. A schematic representation is shown in Figure 4.10.

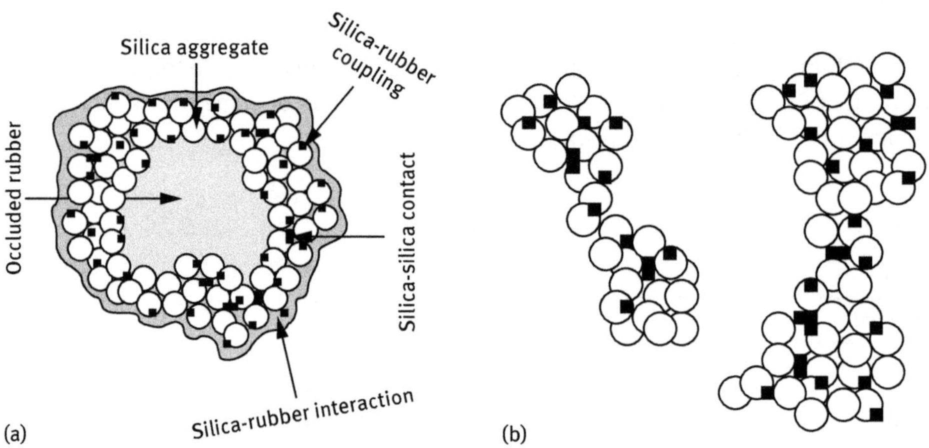

Figure 4.10: Simple model of silica/silane reinforcement: (a) no deformation and (b) after large deformation.

The large polarity difference between the silica filler and the rubber matrix results in the easy formation of a filler–filler network, which also implies that part of the rubber matrix is occluded in the filler network. This means that the occluded rubber is physically and chemically immobilised within the filler network. However, under high deformation, this filler network partially breaks and, with increasing deformation of the rubber, the occluded rubber within the filler network is reduced followed by matrix deformation. Due to chemical bonding *via* a silane coupling agent, the occluded rubber and rubber on the silica surface remain immobilised, and therefore still contribute to the modulus even at high deformations. This chemically immobilised rubber is defined as 'in-rubber structure'.

4.7.6 Use of modified solution styrene-butadiene rubber

Depending on the degree of interaction between the polymer and silica, bound rubber in silica-filled rubber plays an important role in viscoelastic properties. Block copolymers and functionalised polymers are generally synthesised by living anionic polymerisation. Recently, solution-styrene butadiene rubber (S-SBR), a random copolymer synthesised by living anionic polymerisation with Sn-coupling or a functionalised terminal group, has been commonly used to improve the polymer–filler interaction. These functionalised S-SBR basically have polar groups that can directly react with silica [36, 37].

4.8 Technical difficulties in using silica

Silica fillers are not very compatible with non-polar rubber in various respects. Therefore, mixing of silica filler with rubber is a challenge.

4.8.1 Surface energy

In general, surface free energy consists of a 'dispersive part' and a 'specific part', as mentioned earlier. In the case of silica fillers, the 'dispersive part' is low, resulting in a weak interaction between the filler particles and rubber, so the reinforcing effect is also low. However, the 'specific part' is high, resulting in a strong filler–filler interaction, so the viscosity of silica compounds increases substantially. The interparticle forces between filler particles need to be overcome during mixing so that a satisfactory dispersion of filler particles can take place in the retread compound and a link between the silica particles and rubber be established. To solve this, a coupling agent such as bis(triethoxysilylpropyl)disulfane (TESPD) can be used, which can serve both purposes. With a coupling agent the viscosity of the silica-filled compound is also reduced, and the reinforcing effect of the filler also increases as the interaction between filler particles and rubber increases [16].

4.8.2 Solubility

In accordance with a homogeneous distribution of filler aggregates within the matrix, intensive contact between filler particles and rubber is also a necessary condition for good mixing. The surface area of the filler and its wettability are the two most important characteristics determining this contact between the filler surface and rubber matrix [38]. Wettability is determined by the difference in the solubility parameters of the two components. In the case of silica fillers, wettability is quite significant, as silica fillers are polar in nature, which affects their compatibility with non-polar elastomers as well as the cure characteristics.

4.8.3 Structure of the filler

Silica filler has 'high structure'. The aggregates and agglomerates of silica filler are characterised by a high surface area and a high percentage of void volume within the structure of the filler. Polymer chains are physically entrapped in the voids of the filler structure as the direct interaction between the filler and polymer is rather low. The filled structure has to be broken to an optimum level during mixing in order to increase the polymer–filler interaction [17, 39].

The compatibility mismatch between silica fillers and rubber can be overcome by using coupling agents, which can form a chemical linkage between the filler particles and rubber. The coupling agent helps to reduce the specific surface energy and solubility parameter of silica, which greatly improves processing behaviour during extrusion of a retread compound. It also enhances the properties of the final product through the formation of the filler–polymer network [18, 40].

4.9 Silane chemistry

From a reinforcing point of view, carbon black was more efficient than silica for tyre retreads until coupling agents were introduced. The compatibility between silica and polymers is low due to the difference in polarity, so, to make them compatible, this polarity difference must be reduced. This can be achieved by using coupling agents, which are able to react with both the silica surface and the polymer. Coupling agents may be premixed or pre-reacted with the silica particles, or may be added to the rubber mix during mixing, and it is considered that the coupling agents then combine *in situ* with the silica [41].

4.9.1 Types of commonly used coupling agents

To increase the compatibility between the polymer and silica fillers, the polarity difference must be reduced. Silane coupling agents such as bis(triethoxysilylpropyl) tetrasulfide (TESPT), which is capable of reacting with the silica surface and the polymer, are commonly used in silica-filled rubber. The polysulfide part of TESPT reacts with the polymer and the ethoxysilyl groups on the silicon atom react with the hydroxyl groups present on the silica surface, as shown in Figure 4.11.

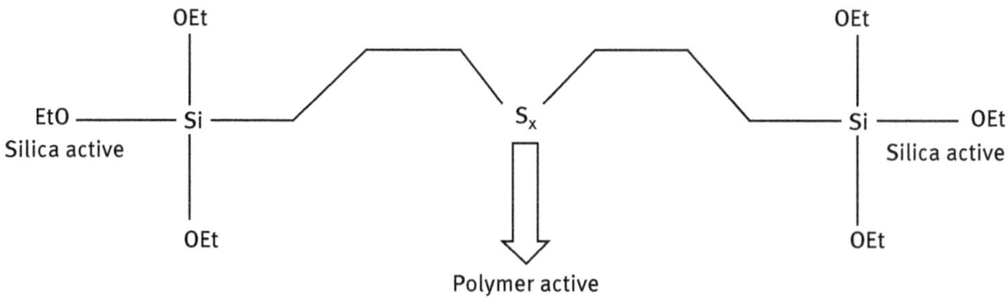

Figure 4.11: Silane coupling agent.

The average sulfur rank of the polysulfide is 3.86. TESPT is unstable at high shear or high temperature. Therefore, during rubber processing, splitting of TESPT molecules takes place, releasing reactive sulfur moieties in silica-filled compounds [42].

An alternative coupling agent, TESPD Si266/Si75, has also been used. TESPD is a mixture of polysulfides rather than a pure disulfide. The average sulfur rank is close to 2. Compared to TESPT, it shows higher stability under high shear conditions as well as high thermal stability. Due to its low sulfur content, elemental sulfur is added from to gain a reinforcement comparable to that of TESPT [43].

4.9.2 Silica–silane reaction

Figure 4.12 illustrates the formation of filler–rubber linkages *via* organosilanes. A bifunctional organosilane coupling agent is generally characterised by its dual functionalities: the silane moiety adheres to the hydrophilic silica surface, and the

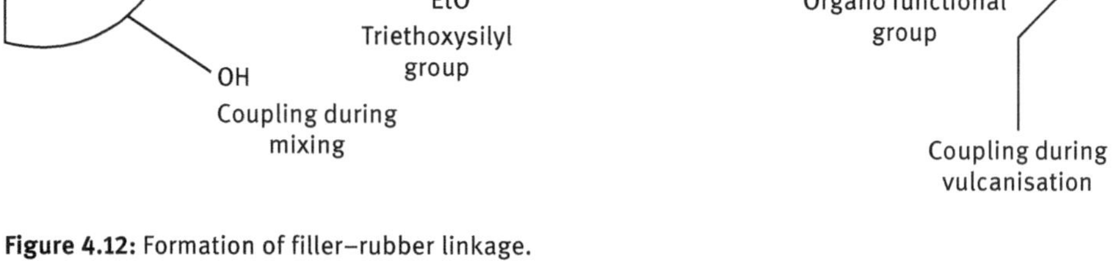

Figure 4.12: Formation of filler–rubber linkage.

other part enhances the compatibility with the hydrophobic polymer matrix [44]. Coupling agents can be applied on the silica particle itself before mixing or to the rubber compound during mixing. Regarding the mixing sequence and mixing temperature, *in situ* modification of silica in the presence of silane coupling agents requires a number of precautions as well as careful adjustment of the mixing conditions. After the addition of coupling agent, due to the formation of a hydrophobic shell around the filler particles the solubility parameter of the filler particles reduces, which results in an increment to compatibility between the filler and rubber phase. This improvement in compatibility also helps during retreading of a tyre tread.

On the other hand, this hydrophobic shell formation also prevents the formation of filler–filler networks as a result of a reduction in the specific surface energy of silica filler [25]. Prior to chemical reaction of the coupling agent with the silanol groups on the silica surface, the silane molecule should be adsorbed on the silica surface [45]. After this, chemical reaction between the silica filler and an alkoxysilyl moiety of the coupling agent takes place in a two-step process:

- *Step 1*: Primary reaction of alkoxy groups with silanol groups on the filler surface. There are two possible mechanisms: (i) direct reaction of the silanol groups with the alkoxy groups of the coupling agent, i.e., direct condensation; or (ii) hydrolysis of the alkoxy groups followed by a condensation reaction with the silanol groups. The rate of silanisation is influenced by the moisture content, which indicates the involvement of a hydrolysis step. The reaction is endothermic in nature and follows pseudo-first-order kinetics [46]. The mechanism of the primary reaction is shown in Figure 4.13.
- *Step 2*: Secondary reaction, a condensation reaction between adjacent pairs of neighbouring silane molecules on the filler surface or between alkoxy groups of the coupling agent and silanol groups of the silica. Compared to the primary reaction, the secondary reaction is slower by a factor of about 10. The activation energy of both primary and secondary reactions is in the range of 30–50 *k*J/mol. The lower value of activation energy indicates a poor temperature dependence of the reaction rate. The reaction is acid- as well as base-catalysed. The rate of the reactions reduces in the presence of sterically hindered groups as well as the electron-donating effects of the leaving group. All the ethoxy groups do not take part during the silanisation reaction. In general, two silane–silica bonds form per cell group, whereas one hydrolysed ethoxy group remains, as depicted in Figure 4.14 [47].

For the reinforcing effect, the coupling agents need a moiety that enables reaction with the polymer during vulcanisation. In general, sulfur moieties – a poly or disulfidic group or a blocked sulfur group – react with the polymer. In other cases, double bonds can be used to link the coupling agents to the polymer. Double bonds have to be activated by the addition of an active sulfur compound or by the

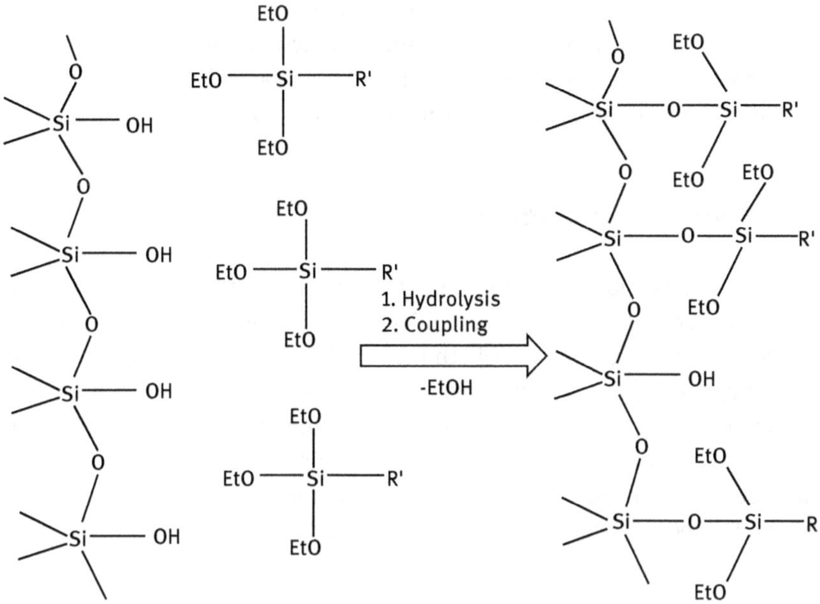

Figure 4.13: Primary reaction of silica with a silane.

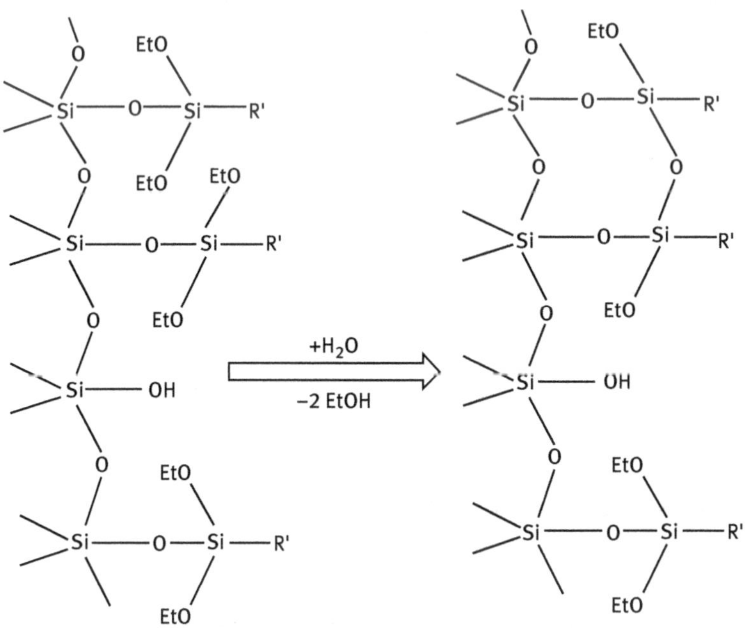

Figure 4.14: Secondary reaction of silica with a silane.

generation of a radical moiety in order to achieve a simultaneous crosslinking of the polymer and the coupling agent with comparable reaction rates during curing.

Cleavage of the unstable polysulfidic moieties of the silanes leads to the formation of active sulfur radicals, which form the filler–silane–polymer bond [48].

4.9.3 Alternative coupling agents

Ethanol formation takes place due to the reaction between silane and silanol groups, as shown in the secondary reaction. Ethanol formation reduces the silanisation efficiency as a result of adsorption of ethanol on the silica surface, and condensation of ethanol in the mixing chamber results in slippage of the compound along the mixing chamber walls, which also reduces the mixing efficiency. To reduce the emission of volatile organic compounds such as ethanol and also to increase the lifetime of retread compounds, new types of silane coupling agents (NXT silanes) has been introduced recently, as shown in Figure 4.15.

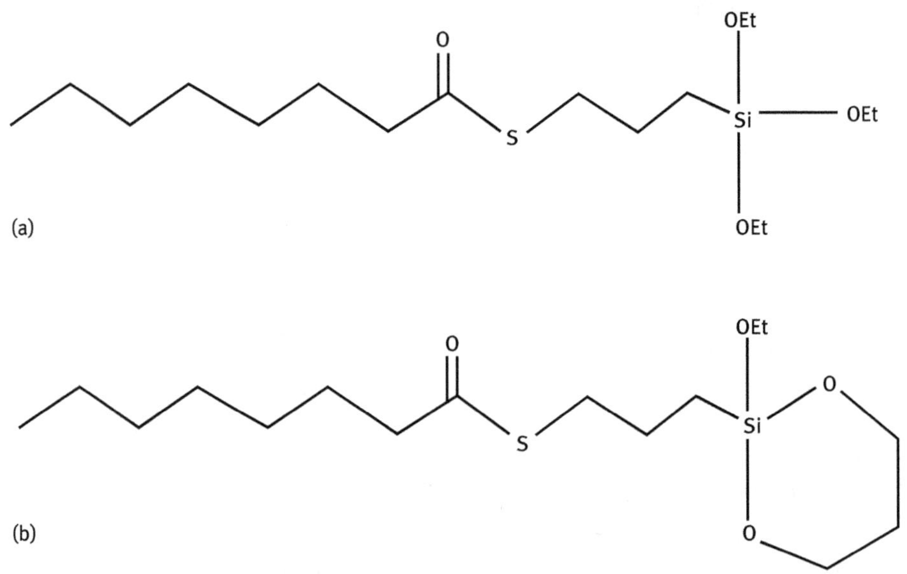

Figure 4.15: New types of silanes: (a) NXT silane and (b) NXT silane containing less volatile organic compounds.

Several NXT silanes have been promoted to alleviate the quantity of volatile organic compounds. However, during processing, NXT silane opposes sulfur donation. By a de-esterification reaction in the presence of an alcohol, the sulfur atom becomes free for a reaction with polymer [49].

4.9.4 Kinetic parameters of the silanisation reaction

To understand the silanisation mechanism, a study related to silanisation kinetics has been carried out [38]. The quantity of ethanol released during rubber processing gives a quantitative estimate of the kinetic parameters of the silanisation reaction. Assuming first-order kinetics, the kinetic parameters of the first-order reaction are estimated by the Equations 4.9 and 4.10:

$$-\frac{\mathrm{d[TESPT]}}{\mathrm{d}t} = k_1[\mathrm{TESPT}] = \frac{\mathrm{d[ETOH]}}{\mathrm{d}t} \tag{4.9}$$

$$\mathrm{In}\ k_1 = \mathrm{In}A - \frac{E_{a1}}{RT} \tag{4.10}$$

where t is time, k_1 is the rate constant of the primary reaction, E_{a1} is the activation energy, R is the gas constant, and T is the absolute temperature [34].

4.9.5 Silanisation acceleration

Silanisation acceleration depends on the rate of hydrolysis of the alkoxysilanes. The rate of hydrolysis is accelerated in the presence of moisture on the silica surface as well as in the presence of an acid or base catalyst. The silanisation reaction, in the presence of an acid or base catalyst, is a bimolecular SN_2-type reaction.

The effect of other silanisation accelerators such as amines, enamines and aldimines has also been studied recently. Use of these amines in the presence of 1,3-diphenylguanidine (DPG) improves the degree of silanisation. In principle, DPG itself acts as an accelerator for the silanisation reaction [50].

4.10 Green tyre technology: The state of the art

According to demand by customers, a good tyre should provide the following features:

- Highest possible traction or cornering force between the tyre and road surface, i.e., good grip, especially under wet conditions.
- Lower rolling resistance, for better fuel economy.
- Good wear or abrasion resistance.

In fulfilling these requirements, the retread compound plays a major role, as tread is the only area in contact with the road surface. In retreaded tyres, the retread compound should fulfil the above criteria. Traction or grip is generally governed by the coefficient of friction between the tread and road surface. A tyre tread must provide high traction and cornering forces between the tread and road surface to create good grip on the road surface and hence avoid slippage. This traction is necessary under both wet and dry road conditions (wet grip and dry grip). Traction predominantly depends on the tyre tread compound, tread profile design, road conditions and tyre construction. Putting aside the construction features, the steering response depends on the tread compound stiffness i.e., the dynamic modulus and friction properties.

Loss of modulus of the retread compound governs rolling resistance. Rolling resistance, being one of the main performance criteria, should be as small as possible with regard to environmental concerns, driving costs and fuel consumption.

To achieve better mileage, the resistance to abrasion of the retread compound should also be as high as possible. The retread compound should show low wear and good durability and also provide optimal driving comfort [51].

Achieving improvements to all three characteristics (rolling resistance, wet grip and tread wear) at once is difficult as they are in conflict. The three performance criteria form the 'magic triangle' of tyre properties shown in Figure 4.16 [52].

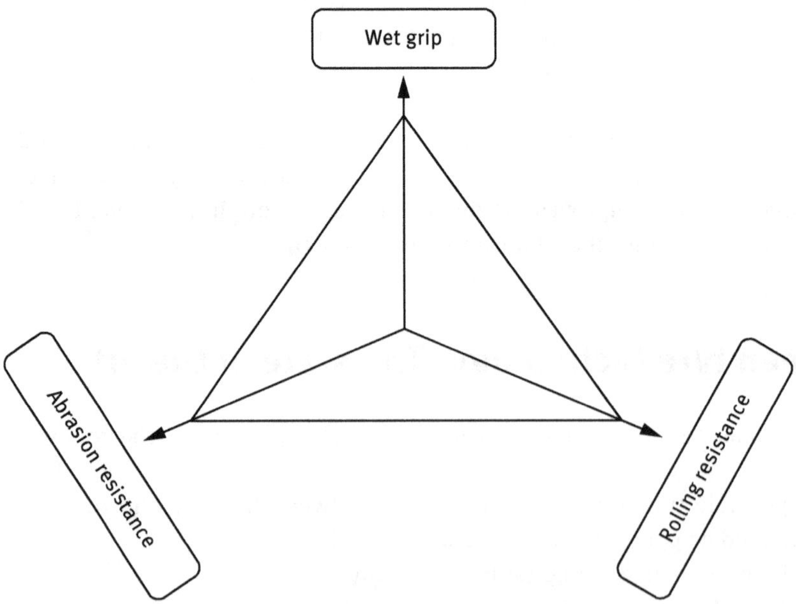

Figure 4.16: Magic triangle of tyre properties.

The concept of 'green tyre technology' means improving road safety and fuel economy and reducing the impact on the environment. In this context, rolling resistance and wet grip are the important matters of concern.

4.10.1 Rolling resistance

From daily experience, most people are very familiar with the phenomenon of rolling resistance. When riding a bicycle, due to resistance to rolling, a substantial amount of effort is needed to keep it moving in the forward direction. Conventionally, rolling resistance has been described as a force acting in the opposite direction of travel, like frictional force, as shown in Figure 4.17.

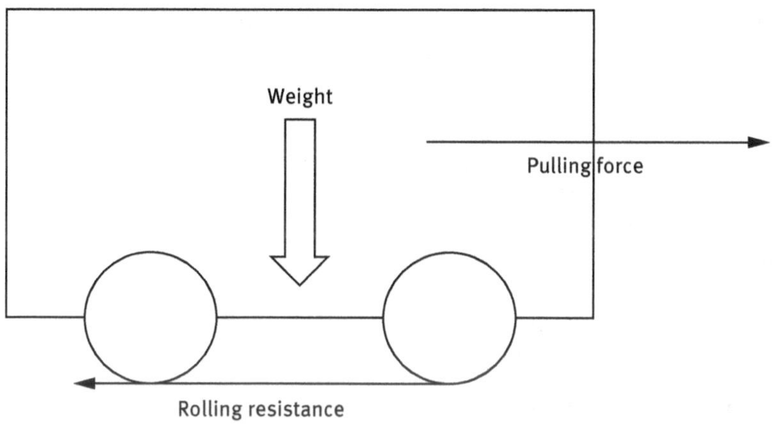

Figure 4.17: Mechanism of rolling resistance.

During rolling of a tyre on a road surface, conversion of mechanical energy into heat energy takes place as a result of the phenomenon called rolling resistance. Rolling resistance is also defined as the energy consumed per unit distance of travel of a tyre rolling under loaded condition [53]. As a force it has the same unit as force (J/m = N), but it is a scalar quantity as it is not associated with any direction. Therefore, rolling resistance F_R can be described by Equation 4.11:

$$F_R = \frac{dH}{dl} \tag{4.11}$$

where dH is the amount of energy converted into heat due to the distance dl travelled. As $dl = vdt$, where v is the velocity of a tyre and dt is the time taken to travel a distance dl, then Equation 4.11 can be converted into Equation 4.12:

$$F_R = \frac{\dot{H}}{V} \tag{4.12}$$

where the heat development rate $\dot{H}$ ($\equiv dH/dt$) is often treated as power loss P_R and is expressed as Equation 4.13:

$$\dot{H} = \frac{dH}{dt} \tag{4.13}$$

Thus, Equation 4.12 can be expressed as Equation 4.14:

$$F_R = \frac{P_R}{V} \tag{4.14}$$

In Equation 4.15, Holt and Wormeley considered the energy balance of the tyre–road wheel system. They proposed that part of the input power delivered by the motor is converted into heat by the tyre, and the rest, the output power, is used to drive the car:

$$P_R = P_{in} - P_{out} \tag{4.15}$$

where P_{in} is the input power of a tyre supplied by the motor, and P_{out} is the output power of a tyre provided for traction of the car. During repeated deformation and friction, the remaining part, P_R, emerges as heating of the tyre. Therefore Equation 4.14 is finally converted to Equation 4.16:

$$F_R = \frac{P_{in} - P_{out}}{V} \tag{4.16}$$

Thus, rolling resistance includes all losses: within the tyre structure, between the tyre and road, and within the road [54].

The tyre therefore consumes a portion of the energy transmitted to the wheels, leaving less energy for moving forward. Only 30% of the consumed fuel is used for the motion of the vehicle in the forward direction, and 70% is lost as heat. Of that 30%, one-third is consumed to overcome aerodynamic drag, one-third for mechanical friction and one-third to overcome rolling resistance, as shown in Figure 4.18.

Rolling resistance plays an important role in raising vehicle fuel consumption. As customer demand for improved road safety, better fuel economy and reduced environmental impact has gained momentum, tyre engineers have searched for affordable solutions. During the last three decades, special attention has been paid to reducing tyre rolling resistance while simultaneously improving other aspects of tyre performance [53]. As a result, tyre engineers are facing the problem of compromising between low rolling resistance, better wet grip and high abrasion resistance.

The lower the rolling resistance, the lower the fuel consumption required to propel the vehicle in the forward direction. However, low rolling resistance results

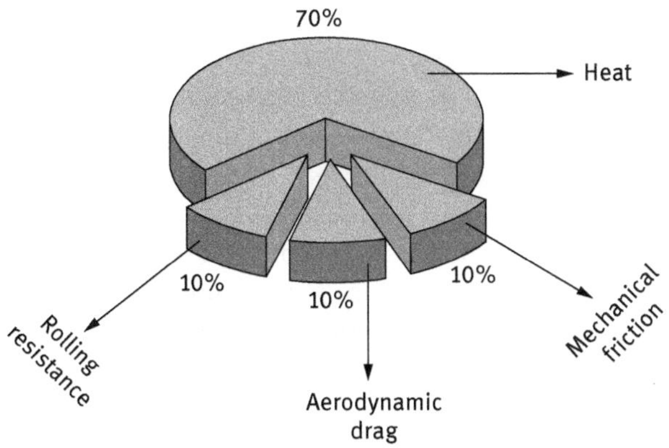

Figure 4.18: Loss of energy during rolling of a tyre.

in poor wet grip, which is unpredictable. For good grip it is necessary to have a high coefficient of friction between the tread and the road surface, but to reduce the rolling resistance and improve fuel consumption, the friction coefficient should be very low. Generally, grip is affected by the degree of distortion of a tyre at high frequencies, and therefore the degree to which it comes into contact with small stones and unevenness in the road surface. Therefore, rubber compounds that absorb a high amount of energy provide the best grip [55].

Rolling resistance, however, is affected by low-frequency distortion. Tread rubber compounds that absorb low quantities of energy (low-hysteresis compounds) provide low rolling resistance. Accordingly, tyre engineers have taken a major step: replacement of carbon black by silica in tyres' tread compound. The incorporation of silica in place of carbon black is also of concern for retreads. The recipe of a retread compound using silica filler is given in Table 4.4 [2, 56].

Table 4.4: Compound formulation.

Ingredients	Quantity (phr)
NR	100
HD Silica	50
Coupling agent	5
Process oil	5
ZnO	4
Stearic acid	1
Trimethyl quinolone	1
DPG	1.2
Cyclohexyl benzthiazyl sulphenamide	1.5
Sulfur	1.5

The addition of silica to retread compounds provides higher hysteresis at high frequencies, giving better wet grip, and lower hysteresis at low frequencies, giving a lower rolling resistance than carbon black [55]. As a result, incorporation of silica in retread compounds produces tyres that can provide good wet grip and lower rolling resistance at the same time. Similarly, in silica-filled retread compounds, the silica filler improves the wet traction of the retread compound with the road surface and also gives a low-rolling-resistance compound that provides lower fuel consumption. The presence of polar groups on the surface of the silica imparts polarity and, as a result of this polarity, silica-filled retread compounds provide better adhesion with the old tyre compound. Silica in retread compounds also gives better properties than carbon black due to a higher reinforcing effect. In the case of 'green tyres', as a result of the addition of silica filler the rolling resistance is reduced by up to 20% compared to an equivalent standard tyre, giving a reduction in vehicle fuel consumption of up to 5% and therefore a reduction in CO_2 emissions. In passenger cars and light trucks a 10% decrease in rolling resistance leads to an improvement in fuel consumption of 0.5–1.5%, while there is a reduction of 1.5–3% in heavy-duty trucks. In 1992 Michelin introduced their green tyre using an S-SBR–silica–silane system, which results in a 20–25% reduction in rolling resistance. According to their estimates, this tyre has saved more than 14.4 billion litres of fuel, resulting in a reduction of 36 million tonnes of CO_2 emissions. Since 2005, Goodyear has introduced silica-filled tread formulations in addition to silane coupling agents, giving a 30% improvement in rolling resistance and saving 9.5 billion litres of fuel [2].

The use of silica also improves wet skid performance. Studies have proved that green tyre traction reduces the stopping distances on wet and icy surfaces by 15%, improving the overall winter driving performance by 10–15% at the same time. After incorporating silica filler into winter tyres, a tyre company has claimed an improvement of wet skid performance of 15%, also substantially improving the braking distances. Compounds containing silica filler, due to their greater elasticity and flexibility at lower temperatures, provide better grip and braking on icy and cold surfaces. Therefore, winter tyre retread compounds are also made with silica, as the silica improves the wet skid resistance, simultaneously improving the rolling resistance [2, 55].

A life cycle assessment by the European Association of the Rubber Industry has also reported that green tyres reduce the impact on human health and environment.

4.10.2 Remedy

Silica has gained more and more importance as a reinforcing filler in tyre applications since the introduction of the 'green tyre'. With coupling agents like TESPT, TESPD or NXT silane, silica forms a good bond with the polymer. This primary reaction is followed by a secondary reaction in which the remaining silanol groups of a coupling agent react with silanol groups of neighbouring coupling agents.

Compared to carbon black, silica has a strong influence on reducing the rolling resistance of tyre tread, which in turn leads to lower fuel consumption. This means that silica reinforcement is of growing importance for the production of energy-efficient tyres in order to preserve the environment and benefit mankind.

The concept of green tyre technology was first proposed in the 1990s, but was not adopted because of the increased costs of manufacturing. Incorporating silica filler into tread compounds on its own leads to poor wear resistance, so although it improves rolling resistance and wet grip, the abrasion resistance becomes poorer than that of carbon black-filled tread compounds. With this in mind, CSDPF has been used by rubber technologists. In other words, the retread compound is not fully reinforced by silica. A combination of carbon and silica filler is now used to achieve the optimum combination of properties in the three aspects of a magic triangle. In fact, CSDPF comprises 90–95% silica filler and 5–10% carbon black. It has been found that these new fillers improve the overall performance of tyre tread in all regards over carbon black or silica filler when used alone. Tyre manufacturers are therefore using CSDPF (2000 series) for truck and bus tread compounds and CSDPF (4000 series) for passenger car tread compounds [57]. In retread compounds, CSDPF is now used to achieve better reinforcing capacity, good wear resistance, low rolling resistance for better fuel economy and good wet grip. By using silica-reinforced retread compounds a tyre can be retreaded several times, although this depends on the type of tyre and also the conditions of use: car tyres, 2–3 times; light truck tyres, 4–5 times; heavy-duty truck tyres, 8–9 times; aircraft tyres, up to 14 times.

References

1. W. Hofmann in *Rubber Technology Handbook*, Hanser/Gardner, Munich, Germany, 1996.
2. J.W. Ten Brinke in *Silica Reinforced Tyre Rubbers*, University of Twente, Enschede, The Netherlands, 2002 [PhD Thesis].
3. C.M. Blow and C. Hepburn in *Rubber Technology and Manufacture*, Plastics and Rubber Institute, London, UK, 1982.
4. J.L. White in *Rubber Processing: Technology, Materials and Principles*, Hanser, Munich, Germany, 1995.
5. S. Wolff, *Rubber Chemistry and Technology*, 1996, **69**, 325.
6. D.J. Zanzig, P.H. Sandstrom, M.J. Crawford, J.J.A. Verthe and C.A. Losey, inventors; The Goodyear Tire & Rubber Company, assignee; EP0638610A1, 1994.
7. M-J. Wang, K. Mahmud, L.J. Murphy and W.J. Patterson, *Kautschuk Gummi Kunststoffe*, 1998, 51, 348.
8. W.K. Dierkes in *Economic Mixing of Silica–Rubber Compounds*, University of Twente, Enschede, The Netherlands, 2005 [PhD Thesis].
9. M.J. Wang and S. Wolff, *Rubber Chemistry Technology*, 1992, **65**, 715.

10 http://www.rtvanderbilt.com/NonBlackFillers.pdf [last accessed 15 July 2012].
11. R.W. Magee, *Rubber .Chemistry and Technology*, 1995, **68**, 590.
12. A. Blume, *Kautschuk Gummi Kunststoffe*, 2000, **53**, 338.
13. W. Kaewsakul in *Silica-Reinforced Natural Rubber for Low Rolling Resistance, Energy-Saving Tires*, University of Twente, Enschede, The Netherlands, 2013 [PhD Thesis].
14. A.R. Payne, *Rubber Chemistry and Technology*, 1966, **39**, 365.
15. A. Einstein, *Annalen der Physik*, 1911, **34**, 591.
16. E. Guth and O. Gold, *Physical Review*, 1938, **53**, 322.
17. H.M. Smallwood, *Journal of Applied Physics*, 1944, **15**, 758.
18. E. Guth, *Journal of Applied Mechanics*, 1945, **16**, 20.
19. G. Kraus, *Rubber Chemistry and Technology*, 1971, **44**, 199.
20. I. Pliskin and N. Tokita, *Journal of Applied Sciences*, 1972, **16**, 473.
21. J. O'Brien, E. Cashell, G.E. Wardell and V.J. McBrierty, *Macromolecules*, 1976, **9**, 653.
22. A.R. Payne and R.E. Whittaker, *Rubber Chemistry and Technology*, 1971, **44**, 440.
23. S. Wolff, A.J. Wang and E.H. Tan, *Rubber Chemistry and Technology*, 1993, **66**, 163.
24. W. Meon, A. Blume, H-D. Luginsland and S. Uhrlandt in *Rubber Compounding Chemistry and Applications*, Ed., B. Rodgers, Markel Dekker, New York, NY, USA, 2004.
25. H.F. Mark in *Rubber Technology Handbook*, Hanser Publishers, Munich, Germany, 1996.
26. A. Blume and S. Uhrlandt in *Proceedings of the 157th ACS Rubber Division Meeting*, Dallas, TX, USA, American Chemical Society, Washington, DC, USA, 2000, Paper No.32.
27. G. Wypych in *Handbook of Fillers*, 2nd Edition, ChemTec Publishing, New York, NY, USA, 1999.
28. L.R. Evans, J.T. Dew, L.T. Hope, T.G. Krivak and W.H. Waddell, *Gummi Fasern Kunststoffe*, 1996, **49**, 456.
29. M.P. Wagner, *Rubber Chemistry and Technology*, 1976, **49**, 703.
30. J. Janzen, *Rubber Chemistry and Technology*, 1971, **44**, 1287.
31. A. Voet, J.C. Morawski and J.B. Donnet, *Rubber Chemistry and Technology*, 1977, **50**, 342.
32. J.L. Leblanc, *Progress in Polymer Science*, 2002, **2**, 627.
33. M.J. Wang, *Rubber Chemistry and Technology*, 1991, **64**, 559.
34. S. Mihara in *Reactive Processing of Silica-Reinforced Tire Rubber*, University of Twente, Enschede, The Netherlands, 2009 [PhD Thesis].
35. H.D. Luginsland, *Rubber Chemistry and Technology*, 2002, **75**, 563.
36. D.W. Sindorf, G.E. Marciel, *Journal of Physical Chemistry*, 1982, **86**, 5208.
37. B. Hsu, A. Halasa, K. Bate, J. Zhou, K-C. Hua and N. Ogata, *International Rubber Conference*, Yokohama, Japan, 26–28th October 2005.
38. W.K. Dierkes and J.W.M. Noordermeer, *Rubber World*, 2004, **229**, 33.
39. J.W. ten Brinke, P.J. van Swaaij, L.A.E.M. Reuvekamp and J.W.M. Noordermeer, *Kautschuk Gummi Kunststoffe*, 2002, **55**, 244.
40. J.E. Mark, B. Erman and F.R. Eirich in *The Science and Technology of Rubber*, 3rd Edition, Elsevier Academic Press, Burlington, MA, USA, 2005.
41. L.A.E.M. Reuvekamp in *Reactive Mixing of Silica and Rubber for Tyres and Engine Mounts*, University of Twente, Enschede, The Netherlands, 2003 [PhD Thesis].
42. S. Bandyopadhyay, P.P. De, D.K. Tripathy and S.K. De, *Rubber Chemistry and Technology*, 1996, **69**, 675.
43. A. Hasse, O. Klockman, A. Wehmeier and H-D. Luginsland, *Kautschuk Gummi Kunststoffe*, 2002, **55**, 236.
44. U. Görl, J. Munzenberg, H.D. Luginsland and A. Muller, *Kautschuk Gummi Kunststoffe*, 1999, **52**, 588.

45. J.G. Matisons, A.E. Jokinen and J.B. Rosenholm, *Journal of Colloid and Interface Science*, 1997, **194**, 263.
46. R.N. Dutta, P.K. Das, S.K. Mondal and D.K. Basu, *Kautschuk Gummi Kunststoffe*, 1988, **41**, 157.
47. M. Marrone, T. Montanari, G. Busca, L. Conzatti, G. Costa, M. Castellano and A. Turturro, *Journal of Physical Chemistry B*, 2004, **108**, 3563.
48. S.C. Debnath, R.N. Datta and J.W.M. Noordermeer, *Rubber Chemistry and Technology*, 2003, **76**, 1311.
49. H-D. Luginsland, A. Hasse, M. Radcziwill and R. Krafczyk, inventors; Evonik Degussa AG, assignee; US1727339, 2004.
50. M.J. Moore, *Rubber and Plastics News*, 2002, **3**, 14.
51. J.E. Mark in *Science and Technology of Rubber*, Academic Press, San Diego, CA, USA, 1994.
52. R. Engehausen, A. Rawlinson and J. Trimbach, *Tire Technology International: Annual Review*, 2001, **36**, 2001.
53. D.E. Hall and J.C. Moreland in *Proceedings of the ACS Rubber Division Meeting*, Dallas, TX, USA, 4–6th April, American Chemical Society, Washington, DC, USA, 2000, Paper No.29.
54. W.L. Holt and P.L. Wormeley, *Technologic Papers of the Bureau of Standards*, 1925, **19**, 213.
55. D.J. Schuring, *Rubber Chemistry and Technology*, 1980, **53**, 600.
56. W.K. Dierkes and J.W.M. Noordermeer, *Rubber World*, 2004, **229**, 33.
57. K. Mahmud, M-J. Wang and R.A. Francis, inventors; Cabot Corporation, assignee; US5830930, 1998.

Dipak Kumar Setua

5 Advances in tyre retreading and scope of run-flat tyres and elastomer nanocomposites

5.1 Introduction

A tyre is truly a composite product, being composed of both metallic components (e.g., wire, made from steel or as brass-coated steel cord) and non-metallic materials (e.g., synthetic and natural rubbers (NR), nylon/polyester/polyvinyl chloride ply, as well as their cord). It also contains fillers such as carbon black, silica, clay, nanoclay and a variety of other additives like sulfur, peroxide, process oil, peptiser, antioxidant, antiozonant, activators (e.g., zinc oxide and stearic acid) and accelerators (*N*-cyclohexyl-2-benzothiazyl sulfenamide, *N*-tertiary butyl 2-benzothiazyl sulfenamide, 2-mercaptobenzthiazole, benzothiazole disulfide, diphenyl guanidine, diorthotolylguanidine and so on). Tyres play an important role in the transportation industry, providing support for vehicles in the form of driving, load bearing, cushioning, vibration damping and transmissibility of torque generated by driving, braking and manoeuvering of the vehicle on the road surface. The tyre also provides the vehicle with superior wet grip and low rolling resistance for driving comfort. The construction of a tyre requires great precision, both in its design as well as in the proper selection and assembly of its constituents (inner liner, casing, bracing, belt plies, carcass, bead wires, sidewalls, tread and so on). The selection of raw materials is also important for a long tyre service life. Tyres are either pneumatic (e.g., those used in bicycles, motorbikes, cars, trucks, off-the-road (OTR) earthmovers, aircraft and so on) or soft solid. More than 50% of the total rubber currently being produced worldwide (NR and synthetic) is used in tyres. The raw materials, additives and synthetic rubbers used in tyre compounding are predominantly petrochemical-based. They therefore consume natural resources, with 55–60% of the total tyre cost going to meet the cost of the raw materials. The function of a tyre includes the dissipation of mechanical work (torque) derived from the energy from the ignition of petrol or diesel in a vehicle engine. Driving also results in heat generation from friction between the tyre and the road. It is therefore necessary that the tyre offer less rolling resistance and thus generate less heat, and also be resistant to deterioration as a result of that heat. This can be accomplished by modifying the tyre

https://doi.org/10.1515/9783110643893-005

tread design and using appropriate compounding materials, which, in turn, also increase the fuel efficiency of the vehicle.

Increasing a tyre's resistance to wear and tear and decreasing abrasion loss have a great impact on service life. Many harmful factors affect the life of a tyre, such as cutting, tearing and groove formation on the tread, ageing due to heat arising from tyre–road friction, rotation of the wheel on sharp turns and at speed breakers, poor vehicle maintenance (e.g., misaligned wheels and inadequate tyre pressure) and poor tread design. Damage to a tread can arise from cutting and chipping, tread detachment, the formation of air pockets, sidewall cut and rupture, bead damage or bursting of the tyre due to over-inflation. A worn-out tyre will become unfit for further use after substantial removal of the tread, and a fully worn-out tread leaves only the bare body, or 'casing'. However, it is possible to make the tyre reusable by subjecting it to some value-added operations. This technology of reactivation of tyres is called 'tyre retreading'. The major benefits arising from this process include cost benefits, reduced vehicle down time, improved safety and a major contribution to keeping the environment clean and green.

The activities involved in reconstructing used tyres include buffing the worn-out tread and bonding a new tread to the casing, a process similar to the manufacture of a new tyre. The process of manufacturing retreaded tyres began in the early 1900s, almost in parallel to the development of new tyres. Tyre retreading is now an established industry. Retreaded tyres should provide safe and dependable vehicle performance at a lower cost than a new tyre. Retreaded tyres have been proven to be dependable, but, for a variety of reasons, the majority of car owners still do not want to use them on their vehicles. Retreading is commonly carried out on truck tyres, because a new truck tyre costs much more than the cost of retreading. Retreading can only be carried out once on car or light truck tyres, but truck and aircraft tyres can be repeatedly retreaded, up to three and twelve times, respectively. However, due care and frequent checks are needed when undertaking repeated retreading for any 'blow-outs' and detachment of the tread from the carcass. Truck drivers are also advised to use new tyres on steering or driving wheels and to use retreaded tyres on the trailers, so that, in the case of an accidental blow-out, the driver does not lose control of the truck. Because the demand for vehicular traffic, trade, military vehicles and so on is growing exponentially, there is a good business potential for the tyre retreading industry.

When a used tyre is received for retreading, a thorough inspection is carried out by an expert. Worn-out tyres are checked visually for damage and rejected immediately if found to be non-retreadable. When there is scope to undertake further repair work, they are subjected to further inspection procedures.

5.2 Visual inspection for external damage

The tyre is first mounted on an inspection spreader and assessed for different types of defects, as described in the following sections.

5.2.1 Bead failure

Beads consist of steel wires that anchor the casing plies and provide a firm mounting surface on the wheel. The failure of beads by bending or breaking occurs as a result of improper tyre mounting, or by demounting, impact damage, improper use of tools, irregular or wrong-sized rims, sudden or abnormal braking of the vehicle or undue acceleration. If the bead is damaged severely then the casing must be rejected. Figures 5.1–5.4 show photographs of several types of bead failure in tyres.

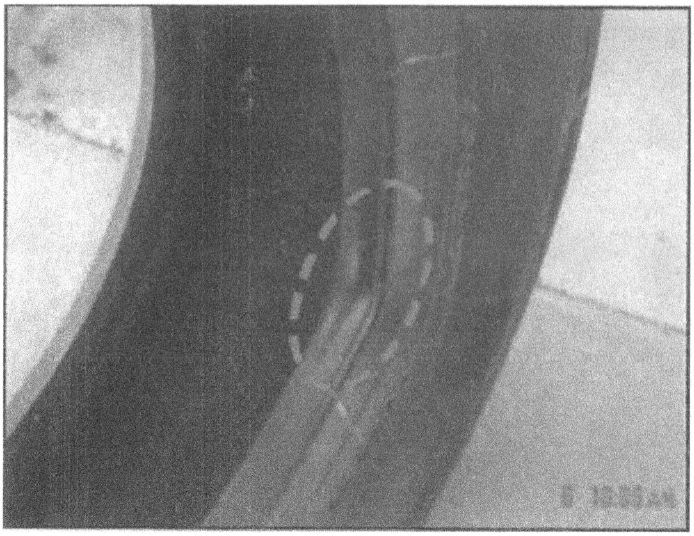

Figure 5.1: Bead area damage due to cone bending. Reproduced with permission from M/S Janatha Rubber Industries, India. ©M/S Janatha Rubber Industries*.

5.2.2 Sidewall failure

The sidewall is the most flexible part of a tyre. Failure of it predominantly arises from the ply cord breaking due to over-inflation, under-inflation, overloading of the vehicle, damage by friction or penetration of any foreign object, and so on.

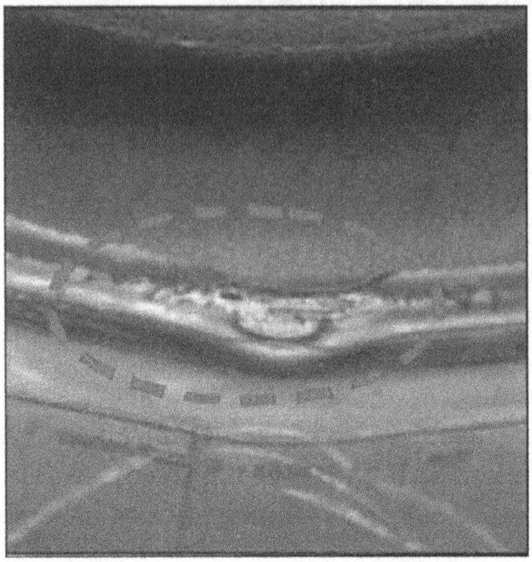

Figure 5.2: Bead area damage due to use of irregular/wrong sized rims. Reproduced with permission from M/S Janatha Rubber Industries, India. ©M/S Janatha Rubber Industries*.

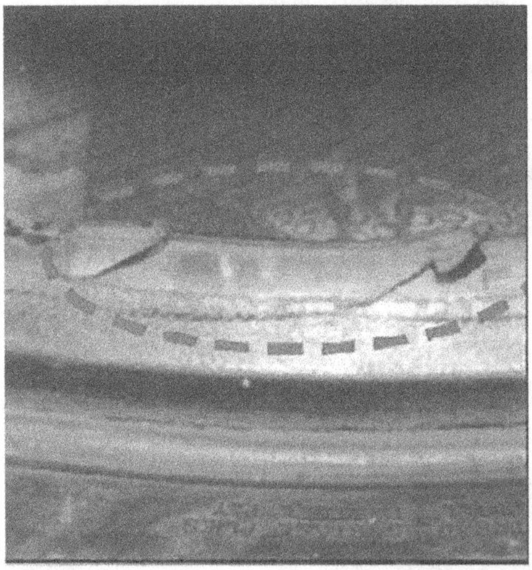

Figure 5.3: Bead area damage due to sudden or abnormal breaking, acceleration, cornering and so on. Reproduced with permission from M/S Janatha Rubber Industries, India. ©M/S Janatha Rubber Industries*.

Figures 5.5 and 5.6 show examples of such damage. Tyres with this level of damage are generally rejected immediately.

5.2.3 Sidewall bulge

This may happen as a result of puncture, impact or inner liner damage, for instance. An example is shown in Figure 5.7. Such types of damage are repairable.

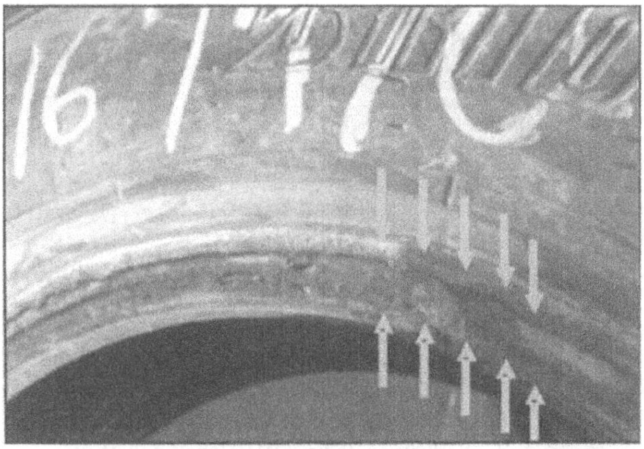

Figure 5.4: Bead damage due to improper mounting or demounting. Reproduced with permission from M/S Janatha Rubber Industries, India. ©M/S Janatha Rubber Industries*.

Figure 5.5: Sidewall cords breaking due to over/under-inflation or overload. Reproduced with permission from M/S Janatha Rubber Industries, India. ©M/S Janatha Rubber Industries*.

5.2.4 Sidewall cut

This may take place as a result of puncture or damage of a tyre by a foreign object. If the rubber is cut and exposes the ply cord only (as in Figure 5.8), it can be repaired. However, if the damage leads to cutting of the cords, the repair must be considered based on the size of the cut. Very deep cutting through the entire ply cannot be repaired and the casing has to be rejected.

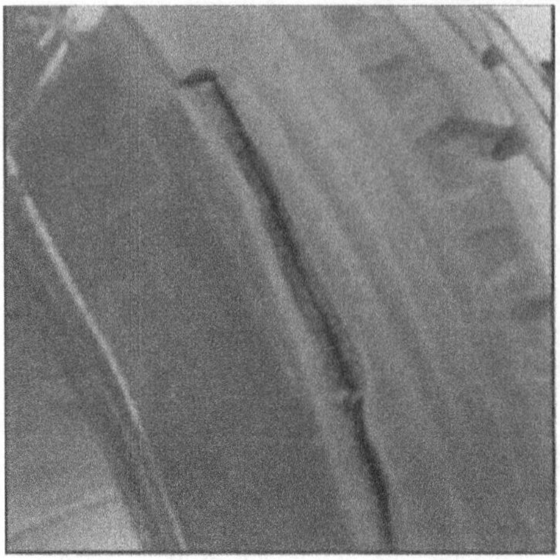

Figure 5.6: Sidewall cords breaking due to damage by a foreign object. Reproduced with permission from M/S Janatha Rubber Industries, India. ©M/S Janatha Rubber Industries*.

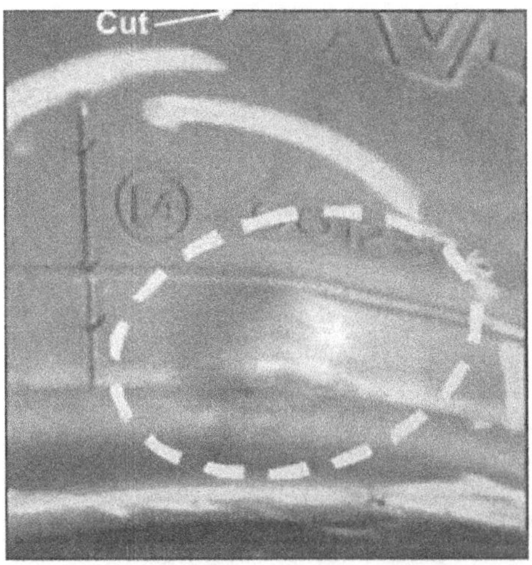

Figure 5.7: Sidewall bulge due to impact or inner liner damage. Reproduced with permission from M/S Janatha Rubber Industries, India. ©M/S Janatha Rubber Industries*.

5.2.5 Sidewall discoloration

This occurs if the tyre continues running after a loss of air due to leakage, puncture, acute under-inflation or over-deflection by overloading. The casings of such tyres generally become weak, and even if retreading takes place, they are not recommended for further use in harsh road conditions.

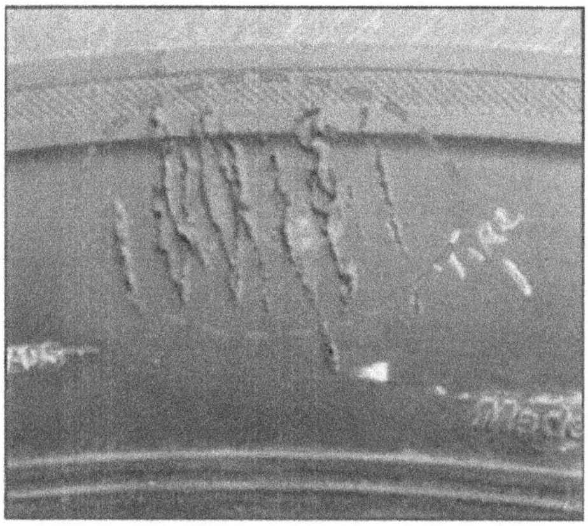

Figure 5.8: Sidewall cut due to piercing by a foreign object or puncture. Reproduced with permission from M/S Janatha Rubber Industries, India. ©M/S Janatha Rubber Industries*.

5.2.6 Ozone cracking

If any small multi-directional crack appears on the sidewall of a tyre, it could be due to prolonged exposure to ozone or sunlight (Figure 5.9). The intensity of ozone cracking is amplified when the tyre is overloaded or underinflated (Figure 5.10). Such tyres are normally very weak and are generally rejected.

Figure 5.9: Sidewall ozone cracking from exposure to sunlight or sources of ozone. Reproduced with permission from M/S Janatha Rubber Industries, India. ©M/S Janatha Rubber Industries*.

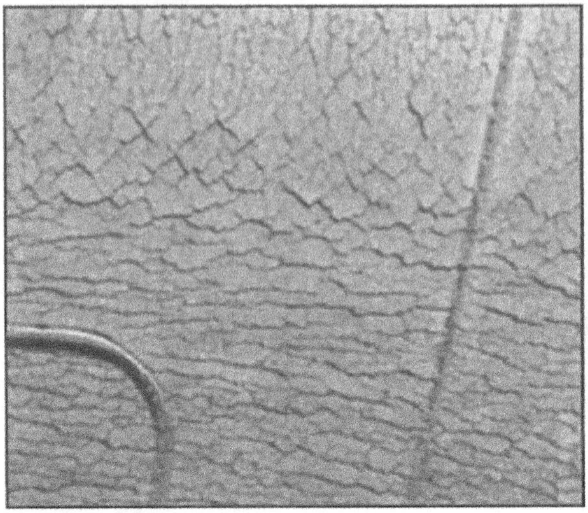

Figure 5.10: Small multi-directional cracks on a sidewall due to overloading, under-inflation or oxidation by ozone. Reproduced with permission from M/S Janatha Rubber Industries, India. ©M/S Janatha Rubber Industries*.

5.2.7 Shoulder damage

In a tyre, maximum heat is generated in the shoulder area and damage can occur for a variety of reasons, including over-deflection as a result of overload, under-inflation, low-quality rubber compounding materials, running at high-speed, or a mismatch of functioning of dual tyres (e.g., in a truck). The end result is separation of the shoulder from the tyre body. If the extent of damage is small, retreading can be performed, but with extensive shoulder damage, the casing is rejected. Figures 5.11 and 5.12 show photographs of some extensively damaged tyre shoulders.

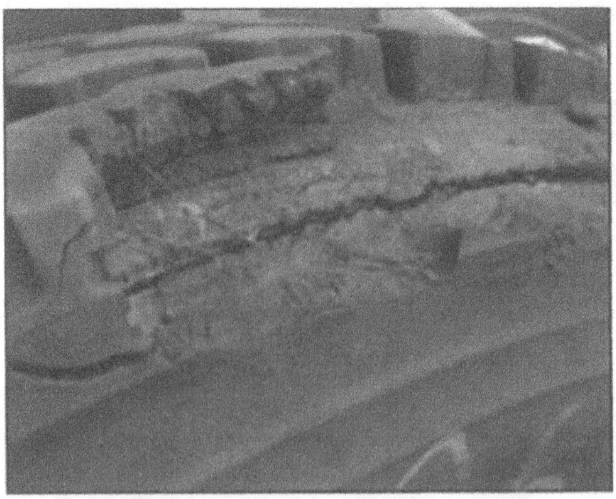

Figure 5.11: Shoulder damage as a result of heat ageing, over-deflection or under-inflation. Reproduced with permission from M/S Janatha Rubber Industries, India. ©M/S Janatha Rubber Industries*.

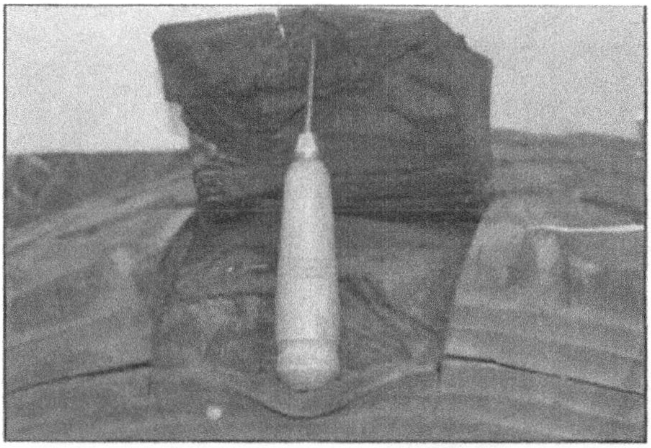

Figure 5.12: Shoulder damage resulting from improper specification, high-speed driving or mismatch between the speeds of dual truck tyres. Reproduced with permission from M/S Janatha Rubber Industries, India. ©M/S Janatha Rubber Industries*.

5.2.8 Crown area damage

Here, damage to the tread is due to penetration by a foreign object (e.g., nails, rocks and so on; Figure 5.13). When the damage extends to the casing and affects the ply, then, based on the prescribed tread wear limiting index, the tyre is rejected for further retreading.

5.2.9 Damage of inner liner

In this case the damage is on the tyre crown, shoulder or sidewall, and penetrates to the inner liner (Figure 5.14). In such cases, the damage zone is tested with a tuning fork. A vibrating sound indicates ply separation or the presence of trapped air. Minor ply separation can be repaired, but when the damage covers a wide area the casing has to be rejected.

5.3 Non-destructive evaluation

Tyres that are visually acceptable for retreading are received, arranged in rows, and stored in a shed for further non-destructive evaluation (NDE) of internal damage. These evaluations are more accurate than visual inspection procedures and consist of several analytical techniques, including X-ray, holography, shearography,

OUTSIDE

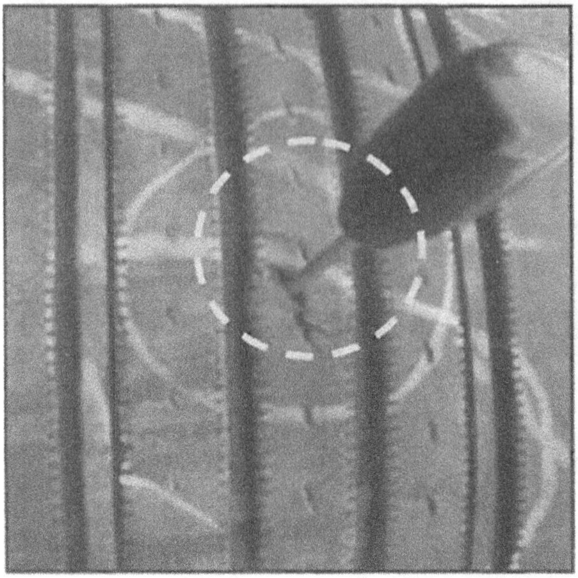

Figure 5.13: Damage to crown area of a tyre by the penetration of a foreign object such as a nail. Reproduced with permission from M/S Janatha Rubber Industries, India. ©M/S Janatha Rubber Industries*.

INSIDE

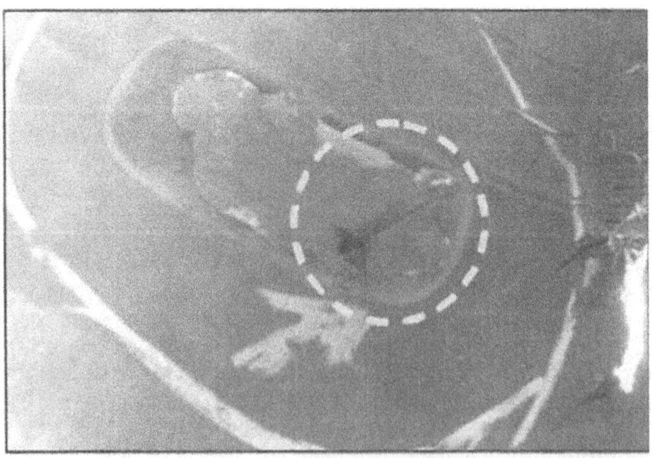

Figure 5.14: Damage of crown area by penetration of a nail up to casing of ply. Reproduced with permission from M/S Janatha Rubber Industries, India. ©M/S Janatha Rubber Industries*.

ultrasound and infrared methods, which have been discussed in detail in Chapter 3 and elsewhere [1]. The radioactive tracer gas technique, an effective NDE procedure, is described in the next section.

5.3.1 Radioactive tracer gas technique

In principle, this technique is similar to the 'Pneutest', which is normally used for quality checking of aircraft and tubeless tyres. The method provides a qualitative assessment of adhesion failure between cord fabrics and the rubber compound. In a new tyre, adhesion is perfect, but a worn tyre will have adhesion defects predominantly at the zones of maximum fatigue. To avoid such a tyre bursting during service it is necessary to detect such defects and take the necessary preventive steps. The NDE technique uses a needle to inject air at 11–12 atm pressure at several locations into the interior carcass of the tyre, above the bead and inside the shoulder (i.e., the regions of maximum fatigue in a tyre). The air is allowed to diffuse out through the carcass. If there is a cord/rubber separation, bulging will occur due to local swelling. This can be seen or measured by the variation in thickness in the region. However, this is rather approximate and unreliable.

In the radioactive tracer gas method, the above technique is modified to make it more accurate. A radioactive gas that emits gamma-rays is injected under pressure in place of air. A mixture of industrial-grade inert nitrogen and the tracer gas ^{133}Xe (chemically inert but biologically safe), with an energy of <100 keV and a half-life of a few days, is administered in a leak-proof manner *via* a needle and allowed to diffuse rapidly within the tyre body. The mixture of gases only accumulates at the defective point. A collimated and highly sensitive probe (a scintillometer or Geiger-Muller Counter, for instance) is kept in contact with the scanned surface around the injection site and the count rate is monitored by a rate meter, with a recorder attached to the probe. With a new tyre the count rate will remain at the background level, even after 10 min of gas injection, but for a used tyre, the carcass is looser, and any adhesion defect between the cord fabric and rubber will result in the accumulation of tracer gas at the failure zones. Defects can also be identified and depicted as a bulge image in a polar scan of the location. Details of the technique, as well as examples and shortcomings, are described in a US patent by Boutaine and co-workers [2]. (In a separate publication, Boutaine [3] has also described quantitative criteria for the acceptance or rejection of aircraft tyres on each landing and successive retreading.) The tyre can be flushed of any residual radioactivity by injecting nitrogen at the original injection site at a higher pressure than used for the initial injection process.

5.4 Safety, green technology, recycling and cost saving in tyre retreading

Retreading is one of the most user-friendly methods with which to mitigate the environmental pollution associated with the disposal of used tyres, which is generally by landfilling, stockpiling or dumping. Retreaded tyres are routinely used on school buses, fire engines, ambulances and military vehicles, and quite often on passenger cars, heavy trucks and construction equipment. A cost–benefit analysis of tyre retreading for light commercial vehicles has been carried out by the Retread Manufacturers Association in the UK, under the Waste & Resources Action Programme (WRAP) [4]. A 10% cost saving is possible by choosing a retreaded tyre instead of a new tyre. If the costs of fuel, wear, insurance and so on are also added, the potential savings can be even greater. This is exemplified by the fact that 98% of the world's commercial airlines and military jets are using retread tyres, and about 80% of aircraft tyres in the US are retread tyres.

In accordance with the global trend, tyre retreading in India has become acceptable, particularly for truck, bus, earthmover and aircraft tyres. However, the share for passenger car tyres remains below expectations. This is partly due to the growth of the market for budget tyres, with the price of a new tyre being only marginally higher than that of a retread. There is also a preference for radial tyres instead of bias tyres (for improved aesthetic appeal) and a desire for new-generation passenger cars. Growing concern for safety is also another reason affecting the sales of retreads. However, retreading for commercial vehicles is poised for higher growth due to the emergence of multi-axle trucks, improved national roadways and connectivity of the road network, as well as the implementation of laws related to restrictions on overloading, traffic control, pollution control and so on. Unlike the practice in developed countries, in India retreading is predominantly carried out using hot/conventional retreading. Indian Standard No. IS 15,724: 2006 covers the automotive tyre retreading procedure following the 'hot process' for passenger cars, light trucks, trucks and buses. With further growth of the economy, more and more new vehicles will emerge and the cold process of retreading (i.e., the precured method) will certainly grow because of potential commercial interest in this sector.

The retreading industry obviously contributes to the conservation of natural resources. Millions of end-of-life tyres are simply thrown into landfill. They do not degrade easily, but continue to pollute the environment. Synthetic rubber, carbon black, rubber chemicals, process oils, tyre cord and so on are all derived from petroleum sources. Retreading is an environmentally friendly process. The Used Tyre Working Group of the Department of Trade and Industry, UK, has reviewed different routes for tyre recycling, such as retreading, recycling, reuse, landfill engineering and energy recovery. They suggest that an average car tyre requires 4.5 gallons less of oil in retreading than required for the manufacture of a new tyre. For a heavy-duty tyre for a commercial vehicle, the savings are even greater (e.g., about 15 gallons of oil) [5]. A new passenger tyre contains rubber and additive equivalent to almost 7–8 gallons of oil, whereas retreading of a worn-out tyre requires only 2–3 gallons, in effect conserving precious oil resources. Pimentel and co-workers [6] have reported on the cost–benefit and environmental advantages associated with the use of retreaded truck tyres. About 30% less energy is consumed in retreading; i.e., an average of 83 litres of oil is required to produce one new truck tyre compared to 56 litres required for retreading.

Dabic and Miljus [7] have discussed the importance of some parameters, such as the number of tyre retreadings, the total distance travelled after each retreading of a tyre up to its writing-off stage and the partial distance traversed by the tyre after each retreading, and some minor factors such as the load/speed of the vehicle, road type, driving style and so on, on the performance of retread tyres for commercial vehicles. As both the collection of used tyres (a time-dependent phenomenon) and the follow-up retreading operations are manually driven activities (which are probabilistic), the whole process becomes stochastic in nature. Mondal and Mukherjee [8] have applied a simulation management strategy to help retreaders make a judicious choice on whether to undertake certain retreading operations or to discard on the basis of cost–benefit concerns. Behnam and Alvelos [9] have explored the benefit of using some quality tools (e.g., Pareto chart, fishbone diagram, matrix diagram and so on) to list the potential causes of tread and ply separation during the retreading process. They have further ranked these causes according to their weight average of importance and identified the critical step to propose effective measures for quality improvement of tyre retreading.

The recycling of waste tyres has also become significant and follows a fast-growing industry trend. However, unlike the recycling of plastics, the market for recycled rubber goods is limited, as the rubber vulcanisates are thermoset and cannot be reprocessed. Tun Abdul Razak Research Centre, at the behest of WRAP, has evolved a surface polymerisation technique that allows a high percentage of rubber crumbs from waste tyres to be incorporated into virgin tread rubber compound [10]. The technology comprises the evolution of a specially formulated polymer system that can be surface-deposited onto rubber particles to create a new compound. The system allows previously vulcanised materials to be reactivated into a new masterbatch composed of specially formulated low-molecular-weight polymers, part uncured virgin rubber and part waste rubber crumb. The surface polymerisation process can be integrated easily into standard rubber processing equipment and compounding processes. The crumb used for the process is high-grade waste (approximately 30 mesh) from truck tyres that has been separated from any steel and fibre cords. The polymer is specially formulated and mixed with the crumb to give a re-polymerised compound that can be provided in strip or flake form for compounding with the virgin rubber compound at the requisite level. Although these studies focus on the use of re-polymerised crumb at 40 and 50% levels for inclusion in retread compound, the technology is also able to produce rubber compounds containing up to 85% of recycled crumb. Morris [11] has analysed the aspect of energy conservation in recycling *versus* incineration processes for retreading and has also compared energy savings from conventional retreading and from the surface re-polymerisation process. The level of savings lies between 16,200 and 48,800 kJ/kg for retreading with a conventional process, and between 67,000 and 229,000 kJ/kg with the re-polymerisation technique (i.e., a saving greater by up to 148,000 kJ/kg with the latter process).

5.5 Mechanism of rubber failure

Any component or system is considered to have failed when it can no longer fulfil its performance expectations during service. Several processes are involved in tyre failure, including fatigue and hysteresis, abrasive wear, tear, blister formation and smearing. Schallamach [12] was the first to examine the scratches produced by a needle on the surface of NR used in different tyre tread compositions. The simple case of abrasion failure of rubber vulcanisate was qualitatively correlated with its hardness, friction and tear resistance. In cases where the tyres encounter sudden breaking or acceleration, contact with sharp bends or run across speed breakers, for example, the abrasion was found to be greatly affected by the nature of the contact surface, the speed of movement and the evolved heat. An investigation was carried out by Greenwood and Tabor [13] on the friction of tyres on hard spheres and cones (an accelerated laboratory test method, as described in the literature) on a lubricated rubber surface under large deformation. They found that, in the case of spheres, the sliding friction is the same as the rolling friction. Although the interfacial adhesion is small in case of lubricated rubber, the friction loss in such cases was found to be dominated by large deformation of rubber, Young's modulus and hysteresis losses. With conical sliders, however, a direct comparison of rolling and sliding resistances could not be made, but the dependency of friction on deformation and tearing of rubber was observed. Bowden and Tabor [14] have reviewed progress in the understanding of friction, lubrication and the wear behaviour of rubber surfaces. Pal and co-workers [15] have reviewed the correlation between abrasion and other mechanical properties of rubber in the context of the development of rubber nanocomposites for tyre applications. It has been mentioned that the volume loss (V) of rubber due to wear is related to hardness and tensile strength by Equation 5.1:

$$V = C_0 + C_1 \times \text{Shore hardness} + C_2 \times \text{tensile strength} \tag{5.1}$$

C_0, C_1 and C_2 are constants, and for abrasion measured in a commercial abrader, V is given by Equation 5.2:

$$V = k_1 \times (\mu P/\sigma_B) \times L \tag{5.2}$$

where μ is the friction coefficient between rubber and the sliding surface, P is normal load, L is the length of the abrading surface and k_1 is a constant. Parameter σ_B is given by Equation 5.3:

$$\sigma_B = \sigma N \tag{5.3}$$

where σ is the maximum amplitude of the tensile stress and N is the number of cycles. It has also been reported that when the bulk rubber undergoes continuous motion in contact with the abrading wheel of an abrader or on a road surface simulating continuous driving, the dynamic friction decreases with increasing velocity of motion. The friction coefficient and abrasion resistance of filled NR, styrene-butadiene rubber (SBR) and polybutadiene rubber (BR) have also been reported by those authors and were found to be dependent on particle size, surface area, carbon black structure and tensile and tear strengths of the rubber vulcanisate.

In engineering applications, crack growth and failure of a composite material depend on the growth of microcracks formed either inherently during fabrication of the component or generated under repeated stress cycles on prolonged use. Generally, tear is a very important criterion of failure. A higher tear resistance in rubbers manifests as increased toughness and superior ballistic properties. Agarwal and coworkers [16] reviewed the work by several authors on the relationships established about the tearing properties of rubber and rubber-based composites. The viscoelastic behaviour of rubbery materials has been expressed by a characteristic energy that is related to the elastic energy stored in the material at the highly strained zone and at the tip of a growing crack. The tear energy has the relationship given in Equation 5.4:

$$T = d \times E_b \tag{5.4}$$

where E_b is the strain energy per unit volume required to break the specimen, and d is the effective diameter of the small zone of initiation of tear. The tear energy depends on various factors, including methods of testing, nature of the base polymer, temperature, sample geometry and so on. The incorporation of antioxidants has been widely practised for tyre tread applications in order to improve tear strength. Gent and Pulford [17] have reported a two-stage mechanism of tearing for rubber composites – formation of a single tear path followed by its splitting in the case of carbon black-filled rubber. The authors have described the micromechanics of failure in two stages, first the deformation of the specimen, followed by growth of the cut. As little is known about the damaged zone where deformation and fracture take place, the use of scanning electron microscopy (SEM) coupled with energy dispersive X-ray spectroscopy has been found to be an effective tool to provide insight into the nature of failure, the homogeneity of the dispersed materials, the effect of ageing and so on. Setua and co-workers [18–24] and Mukhopadhyay [25] have carried out successful studies of the correlation of tear, abrasion and other physico-mechanical properties of a variety of rubber and rubber-based composites with the topography of the fractured surfaces. Computational texture analysis based on statistical and spectral measures for extracting quantitative features of SEM images of NR vulcanisates subjected to tear fracture has also been reported [26]. Various statistical properties were measured, including the detection of average intensity (mean), average contrast (standard deviation), smoothness, third moment, uniformity and entropy of the surface. The directionality of periodic or almost periodic two-dimensional patterns of an image has been determined using the spectral method. The effect of elevated temperatures on tear resistance of elastomers has also been studied by Setua [27].

Saibel and Tsai [28] have examined the feasibility of an energy balance approach for the wear processes of tyres and have proposed various working models. According to them, the mechanism of wear is interrelated with fatigue and temperature rise during tearing, and is complex in nature as they are related to each other. The extent of wear also depends on the deformation characteristic of rubber and the thermodynamics of the whole process. Rolling resistance is another key performance analysed by the industry, in particular to address environmental concerns (i.e., debris left by the wear of tyre tread in contact with the road surface). A reduction in rolling resistance lowers fuel consumption and is achieved by changing the tread design as well as the rubber compound formulation. Finite element analysis of the rolling resistance of 205/60R15 and 155/70R14 passenger car radial tyres has been reported [29]. The energy of dissipation of the tyres was evaluated by simulation analysis of elastic strain energy using a steady-state rolling simulation available in Abacus software, and the loss tangent was measured with a laboratory visco-analyser. Using a biaxial inflation device, Mott and co-workers [30] have evaluated radial and circumferential strains up to 33% in NR sheet. While the radial strain was found to be independent of position, the circumferential strain monotonically decreases to zero at the edge and there was no discontinuity of strain distribution over the degree of inflation. A finite element model was proposed to fit the position-dependent strain data, which were found to govern the connection between strain energy and crack propagation in rubber, which are significant to fatigue life and failure properties.

In the context of tyre retreading and the mechanics of rubber failure, Smith [31] has proposed eight main reasons for the failure of new and retreaded tyres: (i) casing failure of either the carcass or body of the tyre due to belt separation, sidewall separation or rupture (this could be evaluated using NDE techniques such as X-ray, shearography and radioactive tracer methods); (ii) process failure for a variety of reasons, including bad selection of tyre for retreading, curing failure during retreading due to inadequate adjustment of cure time, pressure and temperature of the autoclave; (iii) incorrect application of tread pattern and size, mostly for application in trucking fleets, onto the casing; (iv) incorrect fleet maintenance and monitoring; (v) driver's negligence on run-over of the tyre on blunt objects, inadequate inflation pressure, overloading, oversight of puncture and so on; (vi) raw material problem, including the use of faulty, expired and low-quality materials; (vii) use of cushion gum stock that has already exceeded its shelf life during shipping or storage; and (viii) poor workmanship of tread build-up and a mould with a deficiency in pattern design, presence of porosity, air pockets, honeycombs not properly buffed or cemented to prevent environmental oxidation or containing excess of mould release greases.

5.6 Elastomeric nanocomposites with special reference to tyre retreading

5.6.1 Nanofillers and their impact on mechanical and functional properties of tyres

Of late, nanofillers have attracted considerable attention from the plastics and rubber industries. This is partly due to their scope to replace conventional fillers (generally used in large quantities) to enhance mechanical properties, reduce product costs and also sometimes introduce new functional properties into composites. However, nanofillers are mostly composed of inorganic materials and differ largely in their composition and structure compared with conventional macro-fillers. Various nanofillers are available, including carbon nanotubes (CNT), activated carbon fibre, natural clay (mined, refined and treated), synthetic clay, natural fibre, silica, layered double hydroxide (LDH) and metallic nanoparticles (e.g., alumina, zinc oxide, titanium dioxide and iron) and nanocalcium carbonate. However, layered silicates (i.e., nanoclay) are the most common and widely accepted nanoparticles in the rubber industry. Using nanofiller in rubber, either alone or in combination with carbon black, can significantly improve mechanical, dynamic mechanical, tribological and thermal properties, which can give high performance characteristics in tyres coupled with very good wet skid and low rolling resistances.

The significant enhancement in the reinforcement of the properties of polymers obtained by the choice of nanofillers rather than micro-fillers can be explained by a comparison of the morphological properties of nanocomposites. The molecular chain lengths of the polymers more closely approach the dimensions of these nanofillers, resulting in an intense intermolecular interaction that influences the macroscopic behaviour of the nanocomposites. The inherent characteristics of the nanofillers (e.g., chemical nature, shape, specific surface area of the particle, average size, orientation and distribution in the polymer matrix, volume fraction, interfacial interaction and bonding ability) are collectively responsible for the generation of superior reinforcement properties for a particular polymeric system. As such, a smaller interparticle distance and high specific surface area of the particles enable close proximity of the matrix for interaction with the nanofillers for superior adhesion and in effect manifestation of substantially improved mechanical and functional properties.

Since the beginning of the 20th century, a wide variety of particulate fillers have been used routinely in the rubber industry for reinforcement, to reduce material costs and to improve the processing properties of rubber compounds. The use of carbon black remains dominant in the production of various rubber products, including tyres, tubes, V-belts, seals and gaskets. The reinforcement properties of carbon blacks are controlled by its particle size, surface area, structure and surface chemistry. There is also some nano-reinforcing effect, as a few carbon particles are always present in the nano-size range that is responsible for greatly influencing the mechanical properties of the designed rubber nanocomposites.

Rubber nanocomposites fall in the same category of organic–inorganic hybrid materials, and the reinforcing components are mostly nanoclay, silica, expanded graphite, CNT, nanofibres and so on. Recently, rubber-clay nanocomposites have drawn attention among researchers and original equipment manufacturers, as the sources of clays are independent of petroleum resources so their use in designing nanocomposites is likely to be environmentally friendly. Both natural and synthetic clays are first modified organically using a cation exchange reaction with organic amines or other chemicals to make them hydrophobic. They are then dispersed in a polymer/elastomer matrix (mostly hydrophobic in nature) by either solution intercalation, *in situ* polymerisation or melt intercalation methods.

Various compatibilisation and dispersion techniques have been adopted to mix nanofillers with different rubbers [32]. Nanoclays have been found to improve the mechanical, thermal and swelling properties of rubbers, and the extent of improvement depends on the morphology, polymer–filler interaction, adsorption/occlusion of polymer on filler surface, cleavage of chemical bonds and energy balance and thermodynamics. Nanoclays drastically improve the tack and green strength properties of rubber compounds in comparison to the addition of hydrocarbon resins (wood resin, coumarone–indene resin and so on). Studies on the tribological characteristics also revealed that nanofillers efficiently arrest the propagation of a smooth tear path and thereby significantly improve the tear and wear resistances of nanocomposites. When using a dual/hybrid filler system (e.g., NR/nanofiller/carbon black), the zeta potential differences between the fillers leads to their closer adherence in a unique interconnected architecture through non-black fillers resting in the nanochannels of small carbon black aggregates. This accounts for the synergism and resultant properties combining the inherent characteristics of the individual fillers. Thermal properties are found to be considerably improved, as well as low rolling resistance, high abrasion resistance and better wet grip, which are of interest to tyre applications. A good correlation between simulation and experimental results for rolling resistance has been predicted by a finite element simulation technique for carboxylated nitrile rubber (XNBR) containing dual fillers (e.g., organoclay/carbon black or organoclay/silica) [29].

Among the methods available for preparing rubber nanocomposites, both latex compounds and melt blending methods are superior in terms of cost, easily controllable processing parameters, established machinery and homogeneous dispersion of nanoclay in varied concentrations in NR, SBR, nitrile rubber (NBR) and XNBR. SEM and transmission electron microscopy (TEM) studies were conducted to obtain morphological information [33]. The comprehensive mechanical performance of these nanocomposites was further supported by percolation theory to explain the high nano-reinforcement effect combined with gas barrier, fatigue resistance and chipping or cutting resistances. The TEM studies also provide evidence of clay layers physically intervening in the propagation of cracks during tearing. Molecular dynamic simulations provide the induced molecular orientation of polymeric chains during the stretching process. Pal and co-workers [15] have reported the result of their studies on epoxidised natural rubber (ENR) and organoclay (Cloisite 20A) nanocomposites prepared by the solution mixing technique. The ENR–clay masterbatch was blended with NR/SBR blends along with an intermediate super abrasion furnace (ISAF) type of carbon black (N234). The morphology of the organoclay incorporated in ENR shows an intercalated structure, while ISAF offered a significant reinforcement effect by interfacing between the SBR and NR phases of the blends. They were further reported to produce high abrasion resistances in tests carried out in DuPont and Deutsche Institüt für Norms abrasion testers.

Different types of clay have also emerged as potential nanofillers. The blending of surface-functionalised clay platelets with a polymer matrix can yield a new class of hybrid materials, commonly known as polymer clay nanocomposites (PCN). These PCN are suited to a wide range of applications in automobile, construction, packaging and biomedical applications. The mechanical and structural properties of nanocomposites prepared by mixing NR latex with an aqueous dispersion of clay, pre-swollen by using water in place of chemical modification by any organic cation exchange, were studied by Rezende and co-workers [34]. The influence of non-rubber components such as phospholipids, proteins and calcium cations (in particular) on gel content, crystallite formation and interfacial compatibility on the mechanical properties of both dialysed and non-dialysed NR latex–clay nanocomposites have been reported by the authors. Nanocomposites of styrene-acrylic latex (polystyrene-*co*-butyl acrylate-*co*-acrylic acid) containing Li, K or Ca of Na-montmorillonite (MMT) have been reported to possess up to a ten-fold increase in the elastic modulus and two-fold increase in the tensile strength depending on the counterion type. Using a quantitative model, it was shown that a strong association between the clay lamellae with the rubber matrix is formed by electrostatic attraction at organic/inorganic interfaces, which is as strong as covalent bonding. Clay platelets were also observed to be well dispersed and eventually form a translucent nanocomposite with enhanced solvent resistance but anisotropic swelling behaviour. NR–clay nanocomposites, prepared by latex mixing, can be used to produce macroscopically homogeneous films with an excellent degree of exfoliation and dispersion of clay in a dialysed rubber. In this case, the effective aspect ratio of the exfoliated clay falls in the range of a single clay lamella, and stiffening of the nanocomposite occurs at a very low critical concentration that is almost an order of magnitude lower than the spherical particles. Thus, enhancement of mechanical properties at high uniaxial deformations is related to ordering of the clay lamella inside the polymer framework, which causes strain-induced crystallisation of the NR chains. For non-dialysed rubber, however, the high ionic strength and ionic exchange phenomena of multivalent cations lead to attraction of the clay lamellae, resulting in agglomeration, a lower degree of exfoliation and inadequate reinforcement and technical properties of the designed nanocomposites.

A NR latex–clay nanocomposite film with a low level of clay loading has been studied for oxygen permeability, which was found to decrease by 66% with 1 phr of nanoclay loading [35]. Jacob and co-workers [36] have also reported the effect of nanoclay interlayer distance on mechanical properties, cure characteristics and swelling resistance. A 50% increase in tensile strength and 150% increase in elongation at break were reported for the nanocomposite with 10% clay concentration. The permeation resistance of the nanocomposite was confirmed by testing the barrier property of gas permeability and fits well with Nielson's model.

Das and co-workers [37] have observed that the extent of both intercalation and exfoliation of clay platelets are increased by using stearic acid mixed with organomodified montmorillonite (OMMT) in place of MMT alone in melt mixing with non-polar (NR, ethylene-propylene diene monomer, BR and so on) as well as polar rubbers (such as NBR, polychloroprene and XNBR). However, the effect was more prominent in non-polar rubbers, with a greater improvement in tensile strength and elongation properties than in polar rubbers. The observation was further supported by the results of analysis with a torque rheometer, morphology analysis by TEM, X-ray diffraction, and the measurement of dynamic mechanical properties by dynamic mechanical analysis (DMA), and so on. Beside stearic acid, the effect of increasing the chain length of the fatty-acids on the extent of intercalation of OMMT in NR was studied by Rooj and co-workers [38]. Docosanoic acid (with 22 carbon atoms) gives rise to the highest interlayer spacing of all the fatty-acids studied. Wide-angle X-ray diffraction, Fourier-Transform infrared spectroscopy, contact angle measurements and so on were conducted to gain supporting evidence for enhanced intercalation assisted by a higher chain length of fatty-acids other than stearic acid. Mechanical properties like tensile strength, modulus and elongation at break were further correlated with the SEM, TEM and X-ray characterisation data. The structure–property relationship of the organoclay-filled NR/BR blends that are of interest to tyre tread applications has been studied by Kim and co-workers [39]. Quite a few reviews on the topic of rubber clay nanocomposites (RCN) have been published recently [40–43]. The emphasis has been on scientific and patent literatures on the preparation, rheology, vulcanisation, barrier, thermal, mechanical and DMA properties, including tyre engineering of the designed RCN. Das and co-workers [44] have reviewed rubber-clay nanocomposites with special reference to degree of dispersion, melt intercalation and exfoliation, chemical modification and masterbatch formation of layered clays in a variety of rubber matrices. Their results on the effect of the addition of OMMT and LDH on the curing, mechanical, thermal and dielectric properties of different rubbers have been reported.

N,N-dimethyldodecylamine (DDT)-modified organic layered silicate (OLS) of MMT-filled SBR has been used to manufacture nanocomposites *via* the latex method. They were further characterised by TEM, X-ray diffraction and so on for filler dispersion and estimation of crosslink density in the presence of DDT [45]. A ternary filler system (e.g., carbon black/silica/OLS) was used to formulate a tyre tread compound. Calcium stearate (at 0.5 phr), in place of stearic acid, was found to be more effective in improving the filler–rubber interaction because of a strong ionic effect between the calcium cation and silicates (with anionic surfaces). The nanocomposites were reported to have a low cost to high performance ratio, excellent tensile strength, superior gas barrier property, and excellent dynamic mechanical properties, chipping/ cutting/wear resistances, flame retardancy, high temperature resistance and very good anti-fatigue properties. The incorporation of layered silicate in solution SBR has been used to develop green tyre compounds with a low rolling resistance and high wet grip for effective application in tyre tread compounds [46]. They have emerged as valuable materials for tyre inner tubes, inner liners and OTR tyre tread, and also show potential for use in tyre tread formulations, replacing major parts of carbon black in tyre formulations with a few parts of layered silicate (e.g., 4–5 phr) to improve fatigue resistance and control the heat build-up effect without greatly sacrificing the end-use properties. These advantages were attributed to clay layers physically preventing the growing cracks from propagating and also inducing molecular orientation during the stretching process [47]. These were clarified by a molecular dynamics simulation method and TEM studies. SBR 1502 (styrene content 23%) filled with 6 phr of organoclay (Cloisite 15A) in association with 25 phr of high abrasion furnace carbon black (N330) has been found to be very effective for the development of tyre tread compound. The compound produces better rolling and abrasion resistances, good wet grip and lowers heat build-up, with improved DMA properties compared to conventional SBR-based tyre tread compound containing 40 phr N330 carbon black alone [48]. These observations are supported by X-ray and TEM analysis. Use of about 5–90 phr layered silicate modified with alkylammonium ions in tyre tread compounds containing conventional plasticiser and a standard sulfur curative package in diene rubbers has been reported by

Heinrich and co-workers [49]. Das and co-workers [50] have described the synergistic effects of an expanded and stearic acid-modified organoclay and carbon black hybrid system in NR-based truck tyre compounds. Stearic acid-modified clay was found to dominate the reinforcing ability over the unmodified version, as found by the determination of mechanical, dynamic mechanical and superior rolling resistance properties. Attapulgite, a natural clay with the structural formula $Mg_5Si_8O_{20}(OH)_2(OH_2)_4.4H_2O$ and fibre-like morphology, shows adsorption of organics on the clay surfaces. When used to develop nanocomposites with NR, it shows enhancement in the curing rate as well as the capability to compatibilise the immiscible blends of NR with other rubbers [51]. The NR–attapulgite nanocomposites can significantly improve tensile strength, wear resistance, thermal stability and solvent resistance. Purified and modified attapulgite has therefore attracted considerable attention for the development of rubber nanocomposites.

The successful use of nanoscale fillers in rubbers can generate improved wet skid resistance, resulting in superior safety, reduced abrasion, as well as providing extended service life to tyres. Among nanofillers, nanoclays are the most commonly used commercial additive for the preparation of nanocomposites, accounting for nearly 80% of the total volume of nanofillers used in the rubber industry. Nanoclays are less expensive than other fillers, and the rubber compounds made with them exhibit an acceptable range of stiffness, toughness, excellent barrier properties, enhanced heat deflection temperature, improved colourability and superior scratch, flame and mar resistances. These particles have a coarser surface due to their higher surface energy, and therefore produce stronger interactions with the rubber matrix, reducing the inner friction and providing better rolling resistance. The nanoparticles also reduce strain vibration, which occurs within tyre materials at high car speeds, and give superior traction, especially on wet roads. Carbon nanofibres, CNT (mainly multi-walled CNT), polyhedral oligomeric silsequioxanes and another polymorph variety of carbon (graphene) are gaining importance. It has been reported that CNT can improve tensile strength (600%), tear strength (250%) and hardness (70%) of SBR in tyre applications [52].

In tyre engineering, the incorporation of rubber nanocomposites into tyre components is of interest because of the higher overall performance, reduced weight and cost, lower rolling resistance, better fuel efficiency, safety and driving comfort, and reduced complexity of construction. Tyre tread performance also depends on (i) lower heat build-up preventing premature tyre failure; (ii) improved traction with the road surface; (iii) better wear and tear resistances; (iv) better air retention; and (v) improvement in aesthetic appeal, for example a choice of colourability/transparency beyond meeting specific requirement (e.g., the use of brominated isobutylene-*co*-paramethylstyrene rubber in the tread for winter traction tyres on icy/asphalt surfaces or on snowy glaciers). Nanosized carbon black, with its extremely low surface area, provides an excellent match between hysteresis and the abrasion resistance of rubber vulcanisate, and is therefore in overwhelming demand for tyre tread. Together with the use of nanoclay or nanosilicas (including organosilane or other coupling agents), the performance of tyre treads is significantly improved, as well as providing a cost benefit, personal safety and satisfactory environmental norms. The rolling resistance of tyres can be improved using a carbon black–silane-treated silica combination, and the stopping distance of a car on wet roads can be reduced by 15–20%, together with a reduction of 5% in fuel consumption. Sanchez and co-workers [53] have reviewed the applications of hybrid organic–inorganic nanocomposites based on rubber clay (vermiculite) and the scope of development of a core–sheath type of structure to improve gas barrier properties in tyres for bicycles, cars and trucks.

For the automobile sector up to 2015, the projected market share for the use of nanomaterials and technologies is 23% [54]. The major automotive tyre producers in the international market are Yokohama Tyre Corp. (Japan), Pirelli S.p.A. (Italy), Goodyear Tire & Rubber Co. (USA), Continental AG (Germany), InMat Inc. (USA), and so on. Elastomeric nanocomposites are used in tyre models such as the Goodyear UltraGrip Ice+, Continental EcoContact5, Michelin Energy Saver and Pirelli Cinturato P1.

5.7 Run-flat tyre

These types of tyre are designed to give temporary relief after an accident (such as deflation as a result of puncture). The vehicle can be driven, at a reduced speed, for a limited period of time depending upon the extent of damage. A run-flat tyre can help the driver by providing the flexibility to maintain his time schedule. The main factors governing the load capacity of a tyre are the size of the air ensemble between the tyre and the wheel and the strength of construction and materials used in making the tyre body to hold the requisite air pressure. Flattening or 'blow-outs' occur due to air leakage, so monitoring air pressure in real time is important. If a tyre can temporarily retain vehicle mobility, even after its air pressure decreases or is lost, this brings great benefit. Three main techniques are used for the construction of run-flat tyres: self-sealing, self-supporting and the use of an auxiliary system [55].

5.7.1 Self-sealing

Rubbers have unique visco-elastic properties, with extensibilities of up to 700% but able to recover their original shape and size on release of an externally applied stress.

Rubber elasticity is provided by macromolecular chains, which are either covalently crosslinked or connected in a network by physical associations as segments of glassy or crystalline domains, ionic aggregates or joined through multiple hydrogen bonds. These crosslinked or physical associations prevent the flow of rubber and restrict the tendency to creep. In a self-healing system, when a tyre is broken or cut it can be repaired by bringing the fractured surfaces together to touch, whereupon they self-heal at room temperature. Repaired portions also regenerate their inherent characteristics of extensibility and reversibility. This behaviour is in striking contrast to conventional crosslinked or thermo-irreversible rubbers, which cannot be repaired to rework after fracture. Design of ditopic and multitopic molecules, which are able to form long-lived chains and show polymer-like behaviour in solutions and in bulk, have been reported by Cordier and co-workers [56]. These molecules are used to develop a supramolecular assembly to generate self-healing and a thermoreversible type of rubber. Another example of a self-healing rubber composition is a functional elastomer consisting of a monomer with one conjugated diene, one vinyl aromatic monomer, plus a functional group of a multidentate ligand of polyamine or polycarboxylic acids bonded to polymer. They are capable of complexing and encapsulating a metal ion partitioned by an adhesive, as reported in a US patent [57]. Self-sealing tyres are designed predominantly to fix punctures in the tread area, and have an extra lining inserted inside the tread area that is coated with a sealant. A seal is first provided around the object when punctured, and the hole is then cemented when the object is removed. The sealing process is very fast, so the loss of air is negligible. Sealants are generally injected through the tyre valve, distributed within and throughout the tyre inner surface during tyre rotation and permanently seal most punctures from nails, bolts or screws up to 3/16 inch in diameter.

5.7.2 Self-supporting

Self-supporting tyres have a rather stiffer internal construction than standard tyres, which is capable of temporarily carrying the weight of the vehicle, even after the tyre loses all its air pressure. These tyres typically have rubber inserts next to or between the layers of heat-resistant cords in the sidewalls to help prevent the cords from breaking in the event of major loss of air pressure. They also contain a special type of bead that allows the tyre to firmly grip the wheels in an air loss situation. Self-supporting run-flat tyres efficiently mask temporary loss-of-air symptoms, so a tyre pressure monitoring system (TPMS), described in Section 5.8, is required to alert the driver. Without a TPMS, the driver may not notice under-inflation and thus not take the necessary action to inflate or repair the tyre at the first available opportunity.

5.7.3 Auxiliary supported run-flat system

This type of run-flat tyre is attached to a special rim containing a support ring made of steel that is connected to the wheel of the tyre. This ring supports the weight of the vehicle on flattening of a tyre and shoulders most of the mechanical tasks done by a standard tyre using the run-flat capability of the wheel.

5.7.4 Commercial makers of run-flat tyre

In 1934, Michelin introduced a tyre for military armoured vehicles with a safety rim inside the tyre that can run on a special foam lining after a puncture. The tyre is also 'semi-bullet proof'. Although its performance is satisfactory, its higher manufacturing cost prohibits its use as run-flats for private automobiles. The commercial market for self-supporting run-flat tyres started in 1935. These tyres had a fabric inner for protection against blow-outs. In 1958, Chrysler, in collaboration with Goodyear Tire and Rubber Company, USA, developed a run-flat tyre using a specialised captive air interlining to carry the weight of the tyre. In 1972, Dunlop launched a combined Denovo 'fail-safe' wheel and tyre system. The company had also evolved the TD/Denloc in 1983 by using the Dunlop Self-Supporting Technology and Run-on-Flat (ROF) tyre. Bridgestone's run-flats use their 'Cooling Fin' technology and are applied to sports utility vehicles (SUVs), minivans and large passenger cars where the high tyre cross-section causes a significant rise in heat in the sidewalls and loss of air on continuous driving. The technology minimises the heat caused by deformation of the reinforced sidewall, with airflow created by the protrusions on the surface of the sidewall facing towards the centre of the wheel cooling the sidewall itself [58]. In the late 1980s, Porsche introduced a run-flat for use in its 959 sports car that does not disintegrate on a test track, even at very high-speeds. The tyre uses a special wheel design to keep the tyre in place when deflated, thus preventing its use other than in specified conditions [59]. Similarly, there are few other models that accommodate run-flat tyres such as Firestone's Run Flat Tyre, the Goodyear Extended Mobility Tyre and ROF, Kumho XRP, Michelin Zero Pressure, Pirelli RFT (Run-Flat Technology) and Yokohama Run-Flat and Zero Pressure System. The International Organization for Standardization has adopted a run-flat extended mobility system featuring an additional support ring attached to the wheel that can support the weight of the vehicle in the event of loss of pressure. Although these systems (e.g., Michelin's PAX) can offer better riding quality because of the sidewall stiffness, which is equivalent to a standard tyre, the requirement for special wheels increases their cost and limits their widespread use.

5.7.5 Performance characteristics and use of run-flat tyres

Depending on their design, some run-flat tyres are superior to standard tyres and improve fuel efficiency, while others show a 20% increase in rolling resistance. Internal bracing in some run-flat varieties reduces deformability. However, due to additional reinforcement in the tyre sidewalls, run-flat tyres usually give a more stable ride. The primary benefit of using run-flat tyres is continued mobility in the case of a sudden loss of air pressure while the vehicle is travelling on highways where a service station may be located far off, thereby allowing time schedules to be maintained.

Self-supporting run-flat tyres have now become common on light trucks and passenger cars in the USA and Europe. Automanufacturers promote these tyres as an alternative to carrying a spare tyre. However, if a tyre is irrepairably damaged by puncture in the sidewall or at the edge of the tread, their repair may be difficult or unsafe. Furthermore, these tyres carry a 20–40% weight penalty over standard tyres. The thicker sidewall also reduces the average vehicle fuel mileage.

Run-flat tyres account for only 1–3% of replacement of total tyre sales, but are likely to pick up with the growth of the luxury car market and their popularity for use in armoured vehicles. A Michelin study released in 2008 found that only 3% of drivers worldwide want run-flat tyres. The US market share is well below 1%. Honda Motor Co., USA, in 2009 introduced run-flat tyres in their Honda Odyssey, Touring and Acura RL models.

5.8 Tyre pressure monitoring system

This is an automatic troubleshooting device that continuously monitors tyre pressure, identifies discrepancies, if any, in the inflation of one or more tyres, and warns the driver with real-time information [60]. Accordingly, sensors are integrated into each wheel (including the spare wheel). The wheel transmits radio signals to a control unit located at the rear of the vehicle and provides information about tyre pressure, temperature and so on. The information is displayed on an onboard computer located in front of the driver's seat. There are two types of TPMS, direct and indirect. The direct monitoring system comprises a pressure sensor located on a valve-mounted stem in the tyre and a wireless transmitter. This gives direct pressure readings. Indirect or band-mounted system sensors are attached to the wheels themselves, where they monitor and compare the wheel speed of each tyre to determine if there is any under-inflation. This system works on wheel speed data; a small tyre will rotate faster than a larger one, and an under-inflated tyre will be smaller than a correctly inflated tyre. These differences are measurable using speed sensors in anti-lock braking system/ electronic stability control systems or by using spectrum analysis with advanced signal processing software for individual wheels. Indirect TPMS systems are more sensitive to the influences of different tyre types and external influences such as road surface and driving speed. TPMS can be fitted to all conventional or run-flat tyres and gives the driver audible and visual radio signals with an individual wheel identification code to ensure that all tyres are properly inflated within the limits of normal fluctuations of pressure and temperature resulting from changes in the environment and driving conditions. Tyres lose inflation over time and require regular checks to maintain the prescribed pressure (between 28–35 psi), as this plays a significant role in providing a stable ride and good vehicle performance, decreased downtime and maintenance, lower fuel consumption as well as extended service life. Continuous use of an under-inflated (more than 25%) tyre will put additional stress on sidewalls. Under-inflated tyres become hotter, and that heat can lead to tyre failure. Tyres will have a reduced life span, and tyre blow-outs, delamination or tread separation can result in accidents.

However, putting TPMS systems in place has financial implications. The batteries (normal life about a decade) of sensors fail and break, especially in extreme climates or under icy and salty conditions. Generally, the batteries are not repairable and the entire sensor has to be changed. There is also normal wear and tear of sensors over time. Furthermore, every time the tyre is replaced, retreaded, balanced, or is in an accident, the TPMS must be newly reprogrammed and reactivated. There is also a lack of standardisation. TPMS can only work with compatible sensors in the tyres. Sometimes, during normal tyre service and routine filling of air, indirect sensors will need to be reset, or the sensor will indicate a bigger tyre on refill and send a false signal to the driver that the tyre is over-inflated. The TPMS is not a stand-alone safety device and does not relieve drivers of the responsibility to check (with a pressure gauge) and maintain the correct tyre pressure, but it reduces the frequency of having to do so.

In all light motor vehicles manufactured after 2007 in the USA, and all new passenger cars manufactured in the European Union from November 2014, it is mandatory to implement TPMS systems (in the US under the Transportation Recall Enhancement, Accountability and Documentation Act) [61]. The introduction of run-flat tyres has made the use of TPMS mandatory, to help in meeting the legal requirement for their use. With a run-flat tyre, the driver does not necessarily notice that the tyre is running flat and needs extra warning. The indirect TPMS system, as well as providing an alarm, advises the driver not to drive beyond the recommended speed of 80 km/h and more than a total distance of 80 km after the puncture.

5.9 New-generation tyres

Bridgestone Corporation has reported the development of its ‘air free concept (non-pneumatic)’ new-generation tyres, which feature improved load-bearing capabilities, low rolling resistance with reduced CO_2 emissions, better driving performance, as well as being environmentally friendly [62]. These tyres have high-strength flexible spokes made of thermoplastic resin. The spokes are stretched towards the inner side of the tyres, supporting the weight of the vehicle and reducing stress and deformation. There is no need to periodically fill these tyres with air, little worry about small punctures, and recyclable materials can be used for the construction of the spokes and tyre tread to meet environmental norms. There is also the concept of the development of 'smart' tyres, using two kinds of chip. A passive chip patched to the tyre transmits information such as identification number, tyre type and date of manufacture. An active chip provides real-time information about temperature and tyre pressure, which may be reprogrammed after each retreading operation. These chips allow improved tyre usage efficiency, continuously evaluating and reviewing operating conditions and assuring that a higher percentage of used tyres are suitable for retreading.

Sumitomo Rubber Industries Ltd, on behalf of Dunlop, has reported a new-generation run-flat tyre that includes a combined-technology tyre profile, an aramid casing, the use of high-thermal-conductivity materials in the construction of the tyre, and a dimpled sidewall. It is claimed that the tyre weight can be reduced, as well as a substantial reduction of stiffness of the tyre, resulting in a driving comfort that is equivalent or better than that of a standard tyre [63]. The American Engineering Group (AEG), USA, and India are working on a joint venture to develop bombproof zero-pressure tyres for pick-up trucks (heavier load vehicle), which continue to function even after gunshots or exposure to improvised explosive devices (IED). The prototype tyre demonstrates better heat dissipation, a robust design and the capability to escort soldiers away from the battlefield to a safe zone with a comfortable ride. Normal run-flats cannot work in an IED situation, but the AEG zero-pressure tyre can still function to enable the vehicle to travel for 30 miles at a speed of 30 m/h [64]. The design of this tyre can be extrapolated for use in tyres made for all-terrain, construction and mining vehicles.

5.10 Summary

The emergence of new models of cars, trucks, buses and other vehicles to cope with the increasing human population and maintain mobility has facilitated tyre retreading practice as a potential commercial and competitive venture. A retreaded tyre costs less than a new tyre, so a user may prefer to use them in order to make savings on ongoing tyre expenditure. The most common process used to apply new tread to a properly buffed tyre is the conventional 'hot cap' process. However, this is appropriate for bias or cross-ply tyres, but not radial tyres, which are becoming increasingly popular in the automobile sector. The precured retreading process, which can be used for radial tyres, requires a separate process that is 30–50%

costlier than the hot cap process. This means that, at least in India, in spite of precured retreaded tyres demonstrating better reliability and enhanced service life, the 'hot cap' process remains prevalent. It is only government organizations, state transport corporations and large transport owners that can afford to absorb the higher costs of precured retreading. However, the preference for and market share of the precured technique are expected to grow with the increasing acceptance by users and improved safety associated with the technology. Adhering to the budgetary constraints of precured retreading, together with the growth in the capability of firms undertaking the job, are important. Aircraft tyres have demanding specifications, but retreaded aircraft tyres are used in abundance in the USA, Europe and the rest of the world, so retreading technology is quite established and meets quality, economy and eco-efficiency norms. Off-road machines such as earth excavators are other big users of retreads. Most of the retread market is composed of large fleet operators and frequent buyers of replacement tyres. Retreading thus provides an extended useful lifetime of a worn-out tyre in terms of kilometres and also contributes fuel efficiency, the maintenance of wet grip and control of external rolling noise.

Rubbers (natural and synthetic) are important engineering materials and have made good progress with the advent of nanotechnology. The introduction of new functionalities while maintaining an affordable cost/benefit ratio is important. In this context, the following major benefits can be derived from using rubber nanocomposites in tyres and in tyre retreading:

- The production of lighter-weight tyres, using much lower loading of nanofillers than conventional macro-sized fillers.
- Substantial improvement in mechanical, thermal, barrier, flame resistance and functional properties (e.g., anti-skid, rolling resistance and abrasion resistance).
- A requirement for lower power input, and thus energy savings, in processing of the rubber compounds compared with those containing macro-sized fillers (e.g., carbon black and silica).
- Rubber nanocomposites also meet safety, environmental and green technology norms.

Carbon black is the most common reinforcing filler used in rubber. However, suitably modified and engineered nanoclays, with their very good gas barrier, thermal resistance and so on, show promise for partial replacement of carbon black or use in combination with it in tyre compounding. Rubber-toughened thermoplastics containing nanoclay are more attractive than conventional rubber–plastic blends for automotive applications. Rubber nanocomposites with nanofillers such as CNT, LDH and metal oxides are less common. However, nanofillers are sometimes used for specific applications. For example, high-surface-area nano zinc oxide (ZnO) is more efficient as an activator and preservative of rubber compounds than conventional ZnO. Studies are also under way regarding the design and development of nanocomposites using a hybrid filler system, generally a combination of carbon black and one or more other nanofillers, to form multifunctional rubber nanocomposites. Despite substantial progress made in the area, further studies are needed to investigate the effect of incorporating curing agents inside the clay galleries of nanoclay to develop *in situ* cured rubber nanocomposites. Further development is also needed to find effective methods for the homogeneous dispersion of nanoparticles into rubbers, understanding the interfacial properties of the rubber–nanoparticle framework, and simulation and modelling tools to correctly predict percentage gain in the final mechanical properties of the designed nanocomposites.

A standard tyre crumples immediately after deflation following a puncture, making it unusable for further driving. A run-flat tyre can maintain the mobility of the vehicle for a short distance with reduced speed, so emergency replacement is avoided. The damaged tyre can then be repaired or replaced at a convenient time. While run-flats offer safety, continuity and confidence to a driver, they do have shortcomings. They invariably require a TPMS, making them 30–40% more expensive than standard tyres. They are not readily available, are difficult to fit onto some types of vehicle, are fuel expensive, and their introduction is not mandatory for tyre manufacturers. The thicker and heavier sidewalls of run-flats also lead to compromises between driving comfort and the necessity to always keep a spare wheel as a reserve because the maximum distance of 80 km over which the inflated tyre can run is comparatively small, and there is the added requirement for a facility to transmit awareness of the tyre puncture to the driver.

5.11 Acknowledgement

Thanks go to Shri Soman Chellappan, Managing Director, Janatha Rubber Industries*, Salem, 636015, TN, India, for kindly providing supplementary literature on the inspection of tyre failure and photographs of damaged tyres.

References

1. N.M. Trivisonno, *Rubber Chemistry and Technology*, 1985, 58, 3, 469.
2. L. Boutaine, G. Courtois and J.C. Tanguy, inventors; Commissariat a l'Energie Atomique, assignee; US3997783, 1974.
3. J. Boutaine in *A Radioactive Tracer Method for NDT of Aircraft Tires*, Defense Technical Information Center, Fort Belvoir, VA, USA, 1974.
4. Information Leaflet: Using Retread Tyres on Light Commercial Vehicles, *The Costs and Benefits of Using Retread Tyres*, Waste & Resources Action Programme (WRAP), Banbury, UK http://www2.wrap.org.uk/downloads/Retreads_Info_Booklet.21deb200.4911.pdf.

5. Environmental Resources Management Factsheet on Tyre Recycling, *Tyre Waste Arising and Recycling Rates*, Department of Trade and Industry, London, UK, 2001. http://www.dti.gov.uk/.
6. D. Pimentel, A. Pleasant, J. Barron, J. Gaudioso, N. Pollock, E. Chae, Y. Kim, A. Lassiter, C. Schiavoni, A. Jackson, M. Lee and A. Eaton, *US Energy Conservation and Efficiency: Benefits and Costs, Environment, Development and Sustainability*, 2004, **6**, 279.
7. S. Dabic and M. Miljus in *Proceedings of the 1st Logistics International Conference*, Belgrade, Serbia, 2013, p.258.
8. S. Mondal and K. Mukherjee in *Simulation of Tyre Retrading Process: An Indian Case Study*, Inderscience Publishers, Onley, UK. http://www.inderscience.com/info/inarticle.php?artid=50161.
9. B. Behnam and H. Alvelos, *International Journal of Engineering Science and Technology*, 2011, **3**, 6, 5337.
10. Tyres Final Project Report No: TYR3-008, *Using Waste Tyre Rubber in Retreads at High Inclusion Rates*, Waste & Resources Action Programme (WRAP), Banbury, UK, 2007. http://www2.wrap.org.uk/downloads/Rubber_recovery_final_version_02_11_07mfh2.f46f75a7.4753.pdf.
11. J. Morris, *Journal of Hazardous Materials*, 1996, **47**, 277.
12. A. Schallamach, *Journal of Polymer Science*, 1952, **9**, 5, 385.
13. J.A. Greenwood and D. Tabor, *Proceedings of Physical Society*, 1958, **71**, 6, 989.
14. F.P. Bowden and D. Tabor, *British Journal of Applied Physics*, 1966, **17**, 12, 1521.
15. K. Pal, S.K. Pal, C.K. Das and J.K. Kim in *Recent Advances in Elastomeric Nanocomposites, Advanced Structured Materials*, Eds., V. Mittal, J.K. Kim and K. Pal, Springer-Verlag, Berlin, Heidelberg, Germany, 2011, **9**, 201.
16. K. Agarwal, D.K. Setua and G.N. Mathur, *Defence Science Journal*, 2002, **52**, 3, 337.
17. A.N. Gent and C.T.R. Pulford, *Journal of Materials Science*, 1984, 19, **11**, 3612 (see also Technical Report No.32 of Office of Naval Research, US Government, 1984).
18. D.K. Setua, S.K. De and B.K. Dhindaw, *Journal of Scanning Electron Microscopy*, 1982, **Part III**, 973.
19. S.K. Chakraborty, D.K. Setua and S.K. De, *Rubber Chemistry and Technology*, 1982, **55**, 5, 1286.
20. D.K. Setua, *Kautschuk Gummi Kunststoffe*, 1984, **37**, 11, 962.
21. D.K. Setua and S.K. De, *Journal of Materials Science*, 1985, **20**, 9, 3375.
22. D.K. Setua, S. Chellapan, A.K. Bhowmick and G.N. Mathur, *Polymer Engineering and Science*, 2002, **42**, 1, 10.
23. K.N. Pandey, D.K. Setua and G.N. Mathur, *Polymer Testing*, 2003, **22**, 3, 353.
24. K. Agarwal, D.K. Setua and K. Sekhar, *Polymer Testing*, 2005, **24**, 6, 781.
25. A. Mukhopadhyay, *Journal of Materials Science and Engineering*, 2013, **A3**, 1, 31.
26. D.K. Setua, R. Awasthi, S. Kumar, M. Prasad and K. Agarwal in *Microscopy, Science, Technology, Applications and Education*, Eds.,Mendez-Vilas and J. Diaz, Formatex Research Centre, Spain, 2010, Microscopy Book Series No.3, 1642.
27. D.K. Setua, *Polymer Communications*, 1984, **25**, November, 345.
28. E. Saibel and C. Tsai in *Tire Wear Model*, Vehicle Systems Research Section, Technical Report No.2, NBS, Washington, DC, USA, 1969.
29. S. Ghosh, R.A. Sengupta and G. Heinrich, *Tire Science and Technology*, 2011, **39**, 3, 210.
30. P.H. Mott, C.M. Roland and S.E. Hassan, *Rubber Chemistry and Technology*, 2003, **76**, 2, 326.
31. D. Smith in *Tyre Failures and Causes*, Retreading Business Ltd, Crewe, UK, 2013. http://www.retreadingbusiness.com/article/tyre-failure-and-causes/.
32. A.K. Bhowmick in *Proceedings of the International Rubber Conference & Exhibition*, Mumbai, India, 2010, Paper No.P-02/3.

33. L.Q. Zhang, Y. Wu, Y. Wang and Y. Lu in *Proceedings of the International Conference on Composite Materials* (ICCM 17), 27–31st July, Edinburgh, UK, 2009. http://www.iccm-central.org/Proceedings/ICCM17proceedings/Themes/Nanocomposites/MULTIFUNC%20NANOCOMP/E3.13%20Zhang.pdf.
34. C.A. Rezende, F.C. Braganca, T.R. Doi, L.T. Lee, F. Galembeck and F. Boue, *Polymer*, 2010, **06**, 026.
35. A. Jacob in *Rubber-Layered Clay Nanocomposites*, Cochin University of Science and Technology, Kochi, India, May, 2008 [PhD Thesis].
36. A. Jacob, P. Kurian and A.S. Aprem, *International Journal of Polymeric Materials and Polymeric Biomaterials*, 2007, **56**, 6, 593.
37. A. Das, K.W. Stockelhuber, R. Jurk, D. Jehnichen and G. Heinrich, *Applied Clay Science*, 2011, **51**, 1, 117.
38. S. Rooj, A. Das, K.W. Stockelhuber, N. Mukhopadhyay, A.R. Bhattacharyya, D. Jehnichen and G. Heinrich, *Applied Clay Science*, 2012, **67**, 6, 50.
39. S.K. Kim, J.H. Kang, Y. Choe and Y.W. Chang, *Macromolecular Research*, 2006, **14**, 2, 187.
40. M. Galimberti in *Advanced Elastomers – Technology, Properties and Applications*, Ed., A. Boczkowska, In Tech, Rijeka, Croatia, 2012, Chapter 4.
41. R. Sengupta, S. Chakraborty, S. Bandyopadhyay, S. Dasgupta, R. Mukhopadhyay, K. Auddy and A.S. Deuri, *Polymer Engineering & Science*, 2007, **47**, 11, 1956.
42. P.M. Visakh, S. Thomas, A.K. Chandra and A.P. Mathew, *Advances in Elastomers II: Composites and Nanocomposites*, Advanced Structural Materials, Volume 2, Springer Link, Bucher, Berlin, Germany, 2013.
43. S. Thomas and R. Stephen in *Rubber Nanocomposites: Preparation, Properties, and Applications*, John Wiley & Sons (Asia) Pte Ltd, Clementi Loop, Singapore, 2010.
44. A. Das, D.Y. Wang, K.W. Stockelhuber, R. Jurk, J. Fritzsche, M. Kluppel and G. Heinrich, *Advances in Polymer Science*, 2011, **239**, 85.
45. W.S. Kim, D.H. Lee, I.J. Kim, M.J. Son, W. Kim and S.G. Cho, *Macromolecular Research*, 2009, **17**, 10, 776.
46. W. Stockelhuber and A. Das in *Rubber Nanocomposite*, Leibniz Institute of Polymer Research Dresden, Dresden, Germany. http://www.ipfdd.de/Rubber-Nanocomposite.463.0.html.
47. Y.P. Wu, W. Zhao and L.Q. Zhang, *Macromolecular Materials and Engineering*, 2006, **291**, 944.
48. J.A. Gopi, S.K. Patel, A.K. Chandra and D.K. Tripathy, *Journal of Polymer Research*, 2011, **18**, 6,1625.
49. G. Heinrich, W. Herrmann, N. Kendziorra, T. Pietag and C. Recker, inventors; Continental Aktiengesellschaftann, Hannover, assignee; US6818693B2, 2004.
50. A. Das, K.W. Stockelhuber, S. Rooj, D.Y. Wang and G. Heinrich, *Kautschuk Gummi Kunststoffe*, 2010, July/August, 296.
51. J. Wang and D. Chen, *Journal of Nanomaterials*, 2013, Article ID 496584. http://www.hindawi.com/journals/jnm/2013/496584/.
52. H. Kantamneni, A. Gollakota and S. Nimmagadda, *Advanced Materials Manufacturing & Characterization*, 2013, **13**, 1, 195.
53. C. Sanchez, B. Julian, P. Belleville and M. Popall, *Journal of Materials Chemistry*, 2005, **15**, 35–36, 3559.
54. *Polymer Nanocomposites Drive Opportunities in the Automotive Sector*, Nanotechnology Spotlight, Nanowerk LLC, Honolulu, HI, USA, 11th January 2012. http://www.nanowerk.com/spotlight/spotid=23934.php.
55. *Tire Tech Information – Run-Flat Tires*, Tire Track, Inc., South Bend, IN, USA. http://www.tirerack.com/tires/tiretech/techpage.jsp?techid=56.

56. P. Cordier, F. Tournilhac, C.S. Ziakovic and L. Leibler, *Nature Letter*, 2008,**451**, 7181, 977.
57. R. Mruk, C.J.M. Kaes and R.F. Roskamp, inventors; The Goodyear Tire and Rubber Company, USA, assignee; US20140148555A1, 2014.
58. *Bridgestone Develops New 'Cooling Fin' Technology for Runflat Tires*, Bridgestone Corp, Nashville, TN, USA. http://www.bridgestone.com/corporate/news/2007080102.html.
59. J.D. Power, *Understanding Run-flat Tires*, Westlake Village, CA, USA. http://autos.jdpower.com/content/consumer-interest/8X9wdjl/understanding-run-flat-tires.htm.
60. *Tire Pressure Monitoring Systems*, CDX Online eTextbook. http://www.cdxetextbook.com/auxil/srssystems/safety/tirepressuremonitoring.html.
61. *How Tire pressure Monitoring Systems Work*, howstuffwork. http://auto.howstuffwork.com/car-driving-safety/safety-regulatory-devices/tire-pressure.
62. *A New Generation of Non-pneumatic Tire Revealed by Bridgestone*, Bridgestone Corp, Nashville, TN, USA. http://www.car-engineer.com/new-generation-non-pneumatic-tire-revealed-bridgestone.
63. *The New Generation Run-Flat Tire*, China International Rubber Industry Exposition, *RubberNews*, 2010. http://www.rubbernews.com/article/20100923/DATA/309239902/the-new-generation.
64. *Bomb-Proof Zero-Pressure Tyre in Development: Armed Forces International News*, Indian Defence Forum, Report Discussion, 18th March 2014. http://defence.pk/threads/bomb-proof-zero-pressure-tyre-in-development.304990/.

Samar Bandyopadhyay

6 Testing and quality assurance for the tyre retreading industry

6.1 Introduction

The analysis of materials is directly concerned with the use of rubber and other raw materials as well as finished retreaded tyre in the tyre retreading industry. It is very important to gain as much information as possible about the material [1]. The analysis of elastomeric materials should be based on comparing the material with a properly chosen control material. Values obtained for controls are often used for quantitative calculations of an unknown value.

Material analysis is also an important element of testing in the tyre retreading industry, whether for quality assurance or developmental activities [2]. The analysis of raw materials generally falls into one of two types: (i) a chemical method or (ii) an instrumental method. For quality assurance testing of raw materials, the results are compared with an available standard specification [3].

The specification of a raw material is based on the requirements or particular application. Such standard specifications for raw materials in the rubber industry are laid down by national and international standards institutions [4]. The preparation of a sample is important in order to obtain a representative specimen. A number of different sample preparation techniques are available, depending on the analysis to be performed [5].

Rubber has unique and wide ranging properties, and it is necessary to understand these before incorporating them into products like tyres, hose, conveyor belts, bridge bearings, shoe sole, cable, mounts, gaskets, seals and rocket insulation [6]. Rubber differs considerably from other engineering materials in that it is the most highly deformable material, exhibiting virtually complete recovery, and it is fairly incompressible with a bulk modulus some thousand times greater than its shear modulus [7].

Two types of testing are discussed in this chapter: (i) chemical analysis and (ii) physical analysis. Both the type of analysis and its importance in the assurance of quality for retreading will be separately discussed in the following sections.

6.2 Chemical analysis

Chemical analysis of materials is the first step in any rubber and allied industry. This helps the user to understand the behaviour of the materials during processing, storing and end-use. Chemical analysis is generally performed for a representative

https://doi.org/10.1515/9783110643893-006

sample taken from the bulk, so the sample preparation technique plays an important role. In general, sample preparation for chemical analysis is done by homogenisation of the material [8–12].

The retreaded rubber product industry also requires chemical tests for quality control, material development purposes, and research and development (R&D) [13]. A number of chemical analyses for the different materials used in the tyre retreading industry are performed to check the quality of the materials.

A summary of the different chemical analyses required to characterise raw materials, their definition, significance and method used, different critical controlling parameters, basic necessary tools for analysis and widely used instrumental techniques [14, 15] is given in Table 6.1.

6.2.1 Chemical analysis: Definition, significance and procedure of measurement

6.2.1.1 Specific gravity: Definition, significance and measurement procedure

This is the ratio between the weight of a unit volume of a vulcanisate and the weight of the same volume of water at a given temperature. It is an important parameter for checking accuracy in compounding and serves as a guide in comparing relative compound costs. There are three major ways to determine the specific gravity: (i) the hydrometer method (used for liquid materials), (ii) the direct weight method (used for solid materials), and (iii) the liquid displacement method (used for polymeric materials).

6.2.1.2 Ash content: Definition, significance and measurement procedure

Ash is the residue obtained after ignition of materials at high temperatures. Ash mainly comprises inorganic impurities such as copper, manganese and so on. These impurities can have serious implications for the properties of final products. A weighed amount of sample is ignited in a muffle furnace at 900 °C (for rubber at 550 °C). Residual material is determined as the ‘ash percentage’.

6.2.1.3 Heat loss: Definition, significance and measurement procedure

This is the weight loss of materials, measured at around 60–70 °C for rubber chemicals and at 125 °C for carbon black. The heat loss of rubber chemicals represents low-boiling-point organic materials present in the sample. For carbon black, the heating loss consists primarily of moisture, but other volatile matter may also be

Table 6.1: Chemical analysis of different raw materials at a glance.

Sl. No.	Raw materials	Chemical analysis	ASTM standards
1	Rubber	Specific gravity	D1298-12b
		Ash (%)	D5667-95 (2010)
		Dirt content of NR (%)	D1278-91a (2011)
		Volatile matter (%)	D5668-09 (2014)
		Nitrogen content (%)	D3533-05
		Stabiliser and bound styrene (%)	D1416-89
		Organic acid and soap (%)	D1416-89
		Oil content (%)	D1416-89
		Micro structure	D3677-10e1
		Copper and manganese content of NR (ppm)	D4075-06 (2012)
2	Carbon black	Specific gravity	D1817-05 (2011)
		Ash (%)	D1506-99 (2013)
		Heat loss (%)	D1509-95 (2012)
		Oil absorption (cc/100 g)	D2414-11
		Pour density (kg/m3)	D1513-05 (2012)
		Iodine adsorption number (g/kg)	D1510-13
		Residue on sieve (%)	D1514-04 (2011)
		Pellet hardness (gmf)	D5230-14
		Tint strength	D3265-11
		Toluene extract (%)	D1618-99 (2011)
		pH	D1512-05 (2012)
		Sulfur content (%)	D1619-11
		Nitrogen surface area (m2/g)	D6556-10
3	Processing oils	Specific gravity	D1298-12b(**)
		Kinematic viscosity (cst)	D445-14e2(**)
		Flash and fire point (°C)	D92-12b(**)
		Pour point (°C)	D97-11(**)
		Aniline point (°C)	D611-12(**)
		API gravity	D1298-12b(**)
		Clay-gel analysis (%)	D2007-11(**)
		Viscosity gravity constant	D2501-14(**)
		Refractive index	D1218-12(**)
4	Antioxidant	Specific gravity	D1817-05 (2011)
		Ash (%)	D4574-06 (2012)
		Heat loss (%)	D4571-06 (2012)
		Softening point (°C)	E28-14 (*) and also by DSC
		Melting point (°C)	D1519-95 (2014)

(continued)

Table 6.1 (continued)

Sl. No.	Raw materials	Chemical analysis	ASTM standards
5	Zinc oxide	Zinc oxide content (%)	D3280-85 (2009)
		Heat loss	D280-01 (2012)
		Acidity (%)	D4569-06 (2012)
		Sieve residue (%)	D4315-94 (2012)
		Total sulfur (%)	D3280-85 (2009)
		Lead and cadmium content (%)	D4075-06 (2012)
		Particle size (nm)	D3037-93
		Nitrogen surface area (m2/g)	D6556-14
6	Stearic acid	Specific gravity	D1817-05 (2011)
		Drop melting point (°C)	D127-08
		Acid number	D1980-87 (1998) (*)
		Iodine number	D1959-97(*)
		Ash (%)	D4574-06 (2012)
		Saponification number	D1962-86 (1996)(*)
		Fe, Ni, Cu, Mn content (ppm)	D4075-06 (2012)
		Titer (°C)	D1982-13
7	Accelerator	Specific gravity	D1817-05 (2011)
		Final melt point (°C)	D1519-95 (2014)
		% Assay of sulfenamide accelerators	D4936-10
		Free mercapto benzothiazole in MBTS (%)	D5044-06 (2012)
		Ash (%)	D5044-06 (2012)
		Heat loss (%)	D4574-06 (2012)
		Solubility/insolubility (%)	D4574-06 (2012)
		Sieve residue (%)	D4934-02 (2012)
8	Retarder	Specific gravity	D1817-05 (2011)
		Ash (%)	D4574-06 (2012)
		Heat loss (%)	D4571-06 (2012)
		Solubility (%)	D4934-02 (2012)
		Melting point (°C)	D1519-95 (2014)
9	Soluble/ insoluble sulfur	Specific gravity	D1817-05 (2011)
		Ash (%)	D4574-06 (2012)
		Heat loss (%)	D4571-06 (2012)
		Oil content (%)	D4573-03 (2014)
		Sieve residue (%)	D4572-89 (2012)
		Acidity (%)	D4569-06 (2012)
		Total sulfur (%)	D4578-06 (2012)
		Insoluble sulfur (%)	D4578-06 (2012)
10	Resin	Ash (%)	D4574-06 (2012)
		Heat loss (%)	D4571-06 (2012)
		Softening point (°C)	E28-14 (*)
		Acid number	D1980-87 (1998) (*)

Table 6.1 (continued)

Sl. No.	Raw materials	Chemical analysis	ASTM standards
		Solubility/insolubility (%)	D4934-02 (2012)
11	Silica	Specific gravity	D1817-05 (2011)
		Heat loss (%)	D6738-11
		pH	D6739-11
		Nitrogen surface area (m2/g)	D1993-03 (2013)
		Residue (%)	D1514-04 (2011)
12	Peptiser	Specific gravity	D1817-05 (2011)
		Ash (%)	D4574-06 (2012)
		Sieve residue (%)	D5461-06 (2012)
		Heat loss (%)	D4571-06 (2012)
13	Adhesion promoter	Specific gravity	D1817-05 (2011)
		Final melting point (°C)	D1519-95 (2014)
		Heat loss (%)	D4571-06 (2012)
		Insolubility/solubility (%)	D4934-02 (2012)
14	Clay	Specific gravity	D1817-05 (2011)
		Heat loss (%)	D4571-06 (2012)
		pH of aqueous extract	D1512-05 (2012)
		Sieve residue (%)	D5461-06 (2012)
		Manganese content (ppm)	D4075-06 (2012)
		Copper content (ppm)	D4075-06 (2012)

Note: All methods are from ASTM Volume No. 09.01
* These methods are form ASTM Volume No. 06.03
** These methods are from ASTM Volume No. 05.01
API: American Petroleum Institute
ASTM: American Society for Testing and Materials
DSC: differential scanning calorimeter
MBTS: Dibenzothiazyl disulfide NR: Natural rubber
Reproduced with permission from *Rubber Science and Technology*, Ed., R. Mukhopadhyay, Shantinath Printing Udyog, Udaipur, Rajasthan, India, 2006. ©2006, Shantinath Printing Udyog [2]

lost. A weighed amount of sample is heated under controlled temperature and the loss in weight is a measure of heat loss.

6.2.1.4 Melting point: Definition, significance and measurement procedure

This is the temperature at which the material changes its phase from solid to liquid under specified conditions. Melting point is an important tool with which to identify materials and to check their purity. It also helps to predict the processing characteristics of materials. Two methods are mainly used: (i) capillary methods and (ii) differential scanning calorimetry.

6.2.1.5 Acidity: Definition, significance and measurement procedure

This test method analyses the acid material, which dissociates in distilled water. Acidity determines the effect of raw materials on the vulcanisation system and on the rate of vulcanisation. Acidity is determined by titration.

6.2.1.6 Insolubility: Definition, significance and measurement procedure

This test method is a general procedure to identify insoluble impurities in suitable organic-solvents. Solubility is an important parameter with which to characterise sulfonamide-type accelerators. Because MBTS is a primary degradation product of sulfenamides, the detection of MBTS is a means of assessing the possible degradation of sulfenamides. Insolubles are a means of determining the MBTS content of sulfenamide. This also indicates the purity of materials. A specimen is dissolved in a prescribed solvent, stirred, and filtered through a Gooch crucible. The insoluble content is calculated from the amount of residue.

6.2.1.7 Oil content: Definition, significance and measurement procedure

This test method determines the amount of hydrocarbon oils added to oil-treated materials. This parameter helps to control fly loss (which occurs when any powdery material is discharged from a height) and proper dispersibility of materials in the rubber matrix. It is measured by extracting the oil with suitable solvents.

6.2.1.8 Softening point: Definition, significance and measurement procedure

The softening point is the temperature at which a disk of sample held within a horizontal ring is forced downward a distance of 25.4 cm under the weight of a steel ball as the sample is heated at 5 °C/min in a glycerin or silicon oil bath. The softening point is generally measured for polymeric materials. It indicates the average molecular weight as well as processing temperature. It also indicates the dispersibility of the materials at the processing temperature. Softening point is measured by the ring and ball method.

6.2.1.9 Acid number: Definition, significance and measurement procedure

This is the number of milligrams of KOH required to neutralise the fatty-acids in 1 gm of sample. This analysis is required to determine the acidity of certain rubber chemicals, which may affect the vulcanising reaction. It is determined using the titrimetric method.

6.2.1.10 Saponification number: Definition, significance and measurement procedure

This is represented as the number of milligrams of KOH reacting with 1 g of sample. It is a measure of the alkali reactive group. A higher saponification number than normal indicates that the fatty-acid has been oxidised. Saponification number is measured by titrating with KOH.

6.2.1.11 Iodine number: Definition, significance and measurement procedure

This is a measure of the number of centigrams of iodine required for 1 g of sample. Iodine number is used to measure the unsaturation in organic materials and provides an indication of the stability of materials. The determination of iodine value is based on the absorption of iodine under conditions selected to promote stoichiometric relations.

6.2.1.12 Toluene discolouration: Definition, significance and measurement procedure

This method describes the degree of toluene discolouration by carbon black extractables. The toluene discolouration value provides an estimate of toluene-soluble discolouring residues present on the carbon black. Toluene discolouration is measured by the transmittance, at 425 nm, of the filtrate obtained from the toluene extract of carbon black, compared to that of pure toluene.

6.2.1.13 Sieve residue: Definition, significance and measurement procedure

This test method describes a procedure for the qualitative evaluation of aggregate size and the aggregate size distribution of powdery chemicals. This test method is used to evaluate the suitability of powdered rubber chemicals for use in rubber compounds that require very small particle clusters to achieve uniform crosslinked networks and is intended to be used to ensure that no excessively large particles are present that would result in network 'flaws'.

A sample of powdered rubber chemicals is wetted with a dilute aqueous solution of defoamer (Triton X-100).The sample is transported by water flow through stacked sieves arranged in order of decreasing mesh size. The material retained on the face of each sieve is dried in an oven. The dry mass of retained material is obtained for each sieve. The percent of retained material is calculated on the basis of the original sample mass.

6.2.1.14 Pour density: Definition, significance and measurement procedure

The pour density of carbon black is defined as the mass per unit volume of pelleted carbon black. This factor is useful to estimate the weight–volume relationship for certain applications, such as automatic batch loading systems, and for estimating the weights of bulk shipments. Pour density is measured by estimating the amount of carbon black (in grams) in a fixed-capacity (624 cm^3) cylindrical container

6.2.1.15 Iodine adsorption: Definition, significance and measurement procedure

This is a measure of the number of grams of iodine adsorbed per kilogram of carbon black under specified conditions. It is related to the surface area and is generally in agreement with the nitrogen surface area. The presence of volatiles, surface porosity or extractables will influence the iodine adsorption number. A weighed sample of carbon black is treated with a portion of standard iodine solution and the mixture shaken and centrifuged. The excess iodine is then treated with standard sodium thiosulfate solution, and the adsorbed iodine is expressed as a fraction of the total mass of carbon black.

6.2.1.16 Surface area: Definition, significance and measurement procedure

This is defined as the area in square metres per gram and indicates the effective area available for interaction with rubber. This is generally measured for fine particles such as carbon black, ZnO and precipitated silica. The higher the surface area, the higher the effectivity of the materials. Surface area is measured by the Brunauer–Emmett–Teller method.

6.2.1.17 Oil absorption: Definition, significance and measurement procedure

This test method determines the oil absorption number of carbon black. The oil absorption number is related to the processing and vulcanisate properties of rubber compounds containing carbon black. It also provides an indication of the structure of carbon black. In this test method, oil is added *via* a constant-rate burette to a sample of carbon black in the mixer chamber of an absorptometer. As the sample absorbs the oil, the mixture changes from a free flowing state to a semi-plastic agglomeration, with an accompanying increase in viscosity. The increased viscosity is transmitted to the torque-sensing system of the absorptometer. When the viscosity reaches a predetermined torque level, the absorptometer and burette will shut off immediately. The volume of oil per unit mass of carbon black is the oil absorption number.

6.2.1.18 Pellet hardness: Definition, significance and measurement procedure

This represents the force required to fracture or crush a carbon black pellet. Pellet hardness is related to several carbon black characteristics, including mass strength and attrition. The subsequent level of dispersion in some mixed compounds containing carbon black may be affected by pellet hardness. Pellet hardness is measured by a pellet hardness tester.

6.2.1.19 Tint strength: Definition, significance and measurement procedure

This is the ratio, expressed as tint units, of the reflectance of a standard paste to a sample paste, both prepared and tested under specified conditions. For broad range of commercial rubber-grade carbon blacks, tint strength is highly dependent on particle size. It can be used as an indication of particle size, but tint strength is also dependent on structure and aggregate size distribution. This means that differences in tint strength within grades of carbon black may reflect differences other than particle size.

A carbon black sample is mixed with a white powder (ZnO) and epoxidised soyabean oil to produce a black-grey paste. This paste is then spread to produce a surface suitable for measuring the reflectance of the mixture with a photo-electric reflectance meter. The reflectance of the tested sample is then compared to the reflectance of industrial tint reference black (ITRB) prepared in the same manner. The tint strength of the sample is expressed in units of reflectance of the ITRB divided by the reflectance of the sample multiplied by 100.

6.2.1.20 Organic acids: Definition, significance and measurement procedure

This method is intended to determine the organic acid remaining in a synthetic rubber. Organic acids in the polymer may affect the cure rate of compounded stock. Thin narrow strips of dried rubber are extracted twice in hot extraction solvent. The solvent extracts are titrated against sodium hydroxide solution using a chosen indicator. The titration and sample mass are used to calculate the organic acids.

6.2.1.21 Nitrogen content: Definition, significance and measurement procedure

This method outlines the procedure to determine the total nitrogen in natural and synthetic rubbers and other nitrogen-containing raw materials. The determination of nitrogen in NR is usually carried out in order to arrive at an estimate of the protein content. This is required to assess the quality of the material and its processing characteristics. This method involves digestion of the rubber with a catalytic mixture followed by distillation from a strongly alkaline solution. The distillate is absorbed in a boric acid solution and the excess base is titrated against the acid.

6.2.1.22 Assay of accelerators: Definition, significance and measurement procedure

Determining the percent active content is known as an assay. This test method is designed to assess the purity of accelerators. These products are used in combination with sulfur for the vulcanisation of rubber. For most commonly used sulfenamide accelerators, a weighed specimen is dissolved in an appropriate solvent, the 'free amine' blank is titrated with standard acid, and the sulfenamide is reduced with H_2S (Equation 6.1):

$$BtSNR_2 + H_2S \rightarrow BtSH + HNR_2 + S \quad (6.1)$$

where Bt is the benzothiazole radical, BtSH is 2-mercaptobenzthiazole, and HNR_2 is free amine.

6.2.1.23 Solvent extractables: Definition, significance and measurement procedure

This is a measure of the amount of material extracted from raw materials by organic-solvents. This method provides a uniform and precise method for the gravimetric determination of organic-solvent extractables of different raw materials. The actual chemical composition of the extract can vary significantly with the nature of materials to be tested. These extractable materials may affect processing, storage as well as curing reaction. A specimen of black is extracted for 48 h in a Soxhlet extractor, the solvent is removed by controlled temperature evaporation, and the extracted residue is determined gravimetrically.

6.2.1.24 pH: Definition, significance and measurement procedure

pH is defined as logarithm of the concentration of hydrogen ions in a medium. In short, it indicates the acidity or basicity of materials. By measuring pH value of the material, one can learn whether the material is acidic or basic. It is better to be aware of the pH because basic materials may accelerate vulcanisation while acidic materials may delay it. A solid sample is placed in a beaker and stirred with distilled water. The whole mixture is then allowed to settle and filtered. The pH of the filtrate is determined using a glass electrode.

6.2.1.25 Soaps: Definition, significance and measurement procedure

This method is intended to measure the soap content in emulsion polymers [styrene-butadiene rubber (SBR)]. Soap is a byproduct of the emulsion process, and may affect the cure rate of the compounded stock as well as tack and adhesion properties. One portion of the solvent extract is titrated with hydrochloric acid using the chosen indicator. From the weight of the original sample and the titrations, the percentage of soap is calculated.

6.2.1.26 Flash and fire point: Definition, significance and measurement procedure

The flash point is the temperature at which the saturated vapour on the surface of oils is momentarily ignited. The fire point is the temperature at which oils ignite and burn continuously. The flash and fire points indicate the process safety of the materials. They also indicate the nature of the processing oils. They are measured using the Cleveland Open Cup apparatus.

6.2.1.27 Pour point: Definition, significance and measurement procedure

The pour point is defined as the temperature at which a frozen oil starts flowing. It identifies the processing characteristics of the rubber processing oils. It is measured by pour point apparatus under specified conditions.

6.2.1.28 Aniline point: Definition, significance and measurement procedure

This is the temperature at which a mixture of the oil and aniline becomes completely miscible. This is an important characterising tool for rubber processing oils. A high-aniline point oil has lower aromaticity and *vice versa*. The aniline point of aromatic oil is 25–50 °C, for naphthenic oil it is 55–75 °C and for paraffinic oil is 100–150 °C. The aniline point also indicates the compatibility of oils with different types of polymer.

Sample oil and aniline are placed in a U-tube, where two separate layers are observed. Both materials are then heated and stirred, and at a particular temperature both materials are miscible to each other. That particular temperature is the aniline point.

6.2.1.29 Viscosity: Definition, significance and measurement procedure

This is the internal friction between layers of fluid as they pass over each other when moving with different velocities, and is a measure of the internal force required per unit area to maintain a unit velocity gradient. Viscosity is important to the processing characteristics of different liquid materials with rubber. It also helps to identify the flowability of materials. A number of methods are available to determine the viscosity of a liquid, but the kinematic viscometer and Brookfield viscometer are widely used.

6.2.1.30 Clay-gel analysis: Definition, significance and measurement procedure

This analysis determines the percent of *n*-pentane insolubles, percent of polar compounds, percent of aromatics and percent of saturates. The concentrations of characteristic hydrocarbon groups determined by this method are used to classify petroleum oil types as used for extending and processing rubbers. Compatibility and certain finished product properties can often be correlated with the composition as determined by this method.

The sample is diluted with *n*-pentane and charged to a glass percolation column containing clay in the upper section and silica gel in the lower section. The *n*-pentane is then charged to the double column until a definite quantity of effluent has been collected. The upper section is removed from the lower section and washed with *n*-pentane, which is discarded. A benzene–acetone (50:50) mixture is then charged to the clay section and a specified volume of effluent collected. From this collection, saturates and polars are measured. Aromatics are calculated by difference, where percent aromatics = [100 – (% saturates + % polar)].

All the analyses described in the preceding section can be controlled by the critical parameters [15] listed in Table 6.2.

Table 6.2: Chemical analysis: critical parameters for controlling.

Sl. No.	Chemical analysis	Critical parameters
1	Specific gravity	Weight of the sample, calibration of the hydrometer, analytical balance, specific gravity bottle
2	Ash content	Sample weight, temperature of muffle furnace, time
3	Heat loss	Sample weight, analysis temperature, time for analysis
4	Sieve residue	Sample weight, type of sieve
5	Melting point	Sample preparation, rate of heating
6	Acidity	Reagent (for analysis) purity, preparation of sample
7	Solubility	Sample weight, temperature, solvent type
8	Oil content	Reagent (for analysis) purity, preparation of sample
9	Bulk attrition	Preparation of sample, speed of mechanical shaker, time
10	Softening point	Sample preparation, rate of heating
11	Acid number	Preparation of sample, purity of reagents
12	Iodine number	Preparation of sample, purity of reagents
13	Saponification number	Preparation of sample, purity of reagents
14	Toluene discoloration	Purity of toluene, preparation of sample, time for extraction.
15	Pour density	Calibration of fixed-volume container
16	Oil absorption	Calibration of absorptometer, preparation of sample, purity of oil
17	Aniline point	Purity of aniline, calibration of thermometer
18	Iodine adsorption	Purity of iodine, calibration of glassware, preparation of sample
19	Pellet hardness	Preparation of sample, analysis method
20	Organic acid	Purity of reagents, time for extraction
21	Nitrogen content	Time for digestion, time for distillation, rate of distillation, purity of reagents
22	Assay of sulfonamide accelerators	Reduction time, purity of reagents, preparation of sample
23	pH	Sample preparation, glass electrode stabilisation
26	Solvent extractables	Purity of solvents, extraction time, rate of heating, preparation of sample
27	Tint strength	Sample preparation, calibration of equipment

(continued)

Table 6.2 (continued)

Sl. No.	Chemical analysis	Critical parameters
28	Soaps	Preparation of sample, time of extraction, purity of reagents
29	Flash and fire points	Calibration of thermometer, preparation of sample
30	Titer value	Calibration of thermometer, preparation of sample, rate of heating
31	Pour point	Rate of heating, purity of sample, calibration of thermometer
32	API gravity	Determination of specific gravity
33	Surface area	Preparation of sample, test method
34	Clay-gel analysis	Purity of reagents, purity of column materials, azobenzene value of clay materials

Reproduced with permission from *Rubber Science and Technology*, Ed., R. Mukhopadhyay, Shantinath Printing Udyog, Udaipur, Rajasthan, India, 2006. ©2006, Shantinath Printing Udyog [2]

The list of basics required in any laboratory that performs analysis of materials is given in Table 6.3.

Table 6.3: Chemical analysis: basic necessary tools for analysis.

Environmental basics	Laboratory benches, preferably with chemical-resistant tops Exhaust hoods Gas for bunsen burners Sinks Air conditioning (where applicable) Heating (where applicable) Lighting Electricity
Safety basics	Fire extinguishers Solvent storage cabinets Reagent storage (away from solvents) Emergency first aid kits Safety showers Eye-wash station Safety glasses for all employees Materials safety data sheets

Table 6.3 (continued)

Equipment basics	pH meter Centrifuge Muffle furnace Bunsen burners Hot plates Stirrer Glassware Ovens Thermometers Microscope Mechanical balance
Strongly recommended basics	Computer Word processing software Glassware drying oven Distilled/deionised water maker Electronic balance Laboratory mill Vice Dissecting tools

Reproduced with permission from *Rubber Science and Technology*, Ed., R. Mukhopadhyay, Shantinath Printing Udyog, Udaipur, Rajasthan, India, 2006. ©2006, Shantinath Printing Udyog [2]

Beside the chemical tests mentioned above, a number of instrumental analyses are also gaining importance in characterising raw materials [13]. The analytical instrumental methods of analysis are found to be most accurate and less tedious. Of the instrumental methods, the spectroscopic techniques, chromatographic and thermal techniques are of increasing value as an analytical tool for qualitative and quantitative analysis. Brief summaries of the principles and applications of different analytical instruments are provided in the following sections.

6.2.2 Thermal analysis

Thermal analysis comprises a series of techniques that measure changes in the physical or reactive properties of materials as a function of temperature and time. It provides information that characterises polymers, organic and inorganic chemicals, and so on.

Thermal analysis systems are sophisticated, yet easy to use, producing thoroughly analysed data quickly and with minimum operator involvement. Broad utility, high reliability and high productivity have made thermal analysis an essential analytical tool for material research and selection, product design, process optimisation and quality control. Two widely used types of thermal analysis equipment are the thermogravimetric analyser (TGA) and the differential scanning calorimeter (DSC).

6.2.2.1 Applications

TGA is used for thermal decomposition, roasting and calcining of minerals, proximate analysis of coal, dehydration, oxidative degradation of polymeric materials, reaction kinetics, and so on. DSC is used to generate phase diagrams and in the analysis of reaction kinetics, dehydration reactions, heats of reaction, heats of polymerisation, purity determination, thermal stability, oxidation stability, glass transition temperature, thermal conductivity, specific heat and so on.

6.2.3 Spectroscopy

Spectroscopy provides one of the most important tools for determining the characteristic properties of a compound. It provides a fingerprint identification and is a powerful tool for the study of molecular structure. A number of spectroscopic techniques are available, of which the following are widely used in the rubber and related industries:
- *Infrared spectroscopy*: This involves examination of the twisting, bending, rotating and vibrational motions of atoms in a molecule.
- *Ultraviolet-visible (UV–vis) spectroscopy*: A given molecular species absorbs radiation only in specific regions of the spectrum where the radiation has the energy required to raise the molecules to an excited state. A plot of absorption *versus* wavelength is called an absorption spectrum of that molecular species and serves as a fingerprint for identification.
- *Atomic absorption spectroscopy*: A combustion flame provides a means of converting analytes in solution to atoms in the vapour phase, free of their chemical surroundings. The free atoms absorb radiation from a source external to the flame. The incident radiation absorbed by the free atoms in moving from the ground state to an excited state provides the analytical data.

6.2.3.1 Applications

Infrared spectroscopy is used for compound identification, quantitative estimation of materials, and to determine purity, UV–vis spectroscopy is used for qualitative and quantitative estimation of materials, and atomic absorption spectroscopy is used for quantitative estimation of the metal content (in ppm).

6.2.4 X-Ray diffraction

X-ray diffraction is gaining in importance because material can be analysed in its solid state. X-ray powder diffraction techniques for the identification of crystalline substances have been applied to a wide variety of analyses with useful results. This technique is also used to characterise crystallite size, orientation and so on. X-ray diffraction methods have the following main advantages: (i) they are non-destructive and (ii) results can often be obtained in less time than by routine chemical analysis.

6.2.4.1 Applications

X-ray diffraction is used for the determination of crystal structure, *cis–trans* isomerism, linkage isomerism, polymer characterisation, particle size estimation, and so on.

6.2.5 Chromatography

In chromatography, two mutually immiscible phases are brought into contact. One phase is stationary and the other is mobile. A sample introduced into a mobile phase is carried along a column containing a distributed stationary phase. Species in the sample undergo repeated interactions between the mobile phase and the stationary phase. At the end of the process, separated components emerge in order of increasing interactions with the stationary phase. The least retarded component emerges first, and the most strongly retained component elutes last.

6.2.5.1 Applications

Chromatography is used to determine material purity, and for material identification and quantitative estimation.

6.3 Physical testing

To characterise green, processed and vulcanised retread products, a number of physical tests are carried out. The tests described in the following sections are of paramount importance for products like tyres, retreaded products, conveyor belts, V-belts, footwear, hose, cable, moulded rubber goods and so on. The different characterising tests for green and cured retread rubber are carried out following different national and international standards.

6.3.1 Preparation of test pieces

A test piece of proper dimensions must be formed prior to each test. Direct moulding can be made from mixed compound. Specimens need to be cut, sliced or buffed from the finished products.

6.3.2 Mixing and moulding

Processing variables can affect the results obtained for a final product to a great extent. Different physical testing is carried out in order to detect the effect of these variables (e.g., state of cure and level of dispersion).

Mixing is carried out in open two-roll mills and/or internal mixers (e.g., Banbury or Intermix types) following standard methods.

The conditions and time of storage between mixing and vulcanisation can affect properties, so it is necessary to store material in a dark and dry atmosphere. The preferred conditioning time is 24 h.

To get a better idea of the correlation between laboratory test results and the data from full-sized factory equipment it is necessary to enforce the tightest possible control over equipment, times, temperatures and procedures.

6.3.3 Cutting/Die out from sheet

The accuracy of the final test result depends considerably on the accuracy with which the test piece was prepared. The first requirement is that the test piece should be dimensionally accurate. It is essential that cutters are very sharp and free from nicks or unevenness in the cutting edge that will produce flaws in the test piece and may produce premature failure. Blunt knives lower the tensile strength on ring test pieces by 5–10%. It is normal to restrict stamping to a sheet no thicker than 4 mm as the 'dishing' effect becomes more severe as the thickness increases. Rotary cutters can be used to produce disks or rings from thin sheet and are necessary for sheet above about 4 mm in thickness to prevent distortion. A lubricant that has no effect on rubber can be applied during cutting the sample, particularly when using a rotating cutter.

6.3.4 Test pieces from finished products

It is obviously desirable to make tests wherever possible on an actual finished product rather than on specially prepared test pieces, which may have been produced under rather different conditions. To obtain a test piece from a finished product it is necessary to cut a large block and then reduce the thickness or remove irregularities by buffing or with a slitting machine. A particular disadvantage of buffing is that heat is generated, which may cause significant degradation of the rubber surface. The effect of buffing on the tensile properties (drop) of soft rubbers is 10–15%, whereas for a tyre tread rubber the drop is about 5–10%. Of the many test methods available to the rubber compounder, only a few of the most common tests, their significance and equipment used will be discussed here (bearing in mind the limited operating budget available to small- and medium-scale manufacturers).

6.3.5 Reasons for physical testing

Physical testing is carried out for quality control of incoming material, quality control of in-process material, quality control of finished goods, factory technical service, customer technical service, the preparation of samples for the sales department, compound development, R&D, for advertisements and for product performance testing. The processability tests (procedures, significance, standard used and so on) carried out in the tyre retreading industry are described [15, 16] in detail in Sections 6.3.5.1–6.3.5.4.

6.3.5.1 Plasticity and recovery test

ASTM D926-08 (2013): Standard test method for rubber property – plasticity and recovery (parallel plate method).

Significance of Testing: This is a very rapid test and needs a very small test piece. The test piece is compressed between two parallel plates under a constant force, and the compressed thickness is measured. The extent of compression gives a measure of the plasticity of the rubber. This test is very useful for routine testing, particularly for raw rubber like natural rubber (NR).

Procedure: ASTM D926 describes the above test based on William's plastimeter (with plates that are 4 cm in diameter). The test piece has a volume of 2.00 ± 0.02 cm^3 and can conveniently be a cylinder that is 16 mm in diameter and 10 mm thick. The test piece is preheated for 15 min at a temperature of 100 °C and compressed under a force of 5 kgf for about 3–10 min.

Measurement and Calculation: The thickness of the compressed test piece is measured in millimeters, and this value is multiplied by 100 to give the plasticity number. After this, the test piece is allowed to cool for 1 min and its height is measured. The 'recovery value' is reported as the difference between the plasticity number and the recovered height, and is intended to be a measure of the elastic component.

Plasticity measurements are made before and after ageing to determine the plasticity retention index (PRI). Test pieces are aged at 140 ± 0.2 °C for 30 ± 0.25

min and cooled to room temperature before making the measurements. PRI is calculated according to Equation 6.2:

$$PRI = (\text{aged rapid plasticity number}/\text{un} - \text{aged rapid plasticity number}) \times 100 \quad (6.2)$$

Limitations: The shear rate encountered in rubber processing operations is very high, at ~10^4 sec^{-1}, but plasticity tests operate at a much lower rate, ~0.0025 to 1 sec^{-1}, and thus have no correlation with injection moulding.

6.3.5.2 Mooney viscosity, stress relaxation, mooney scorch

ASTM D1646-96: Standard test methods for rubber – viscosity, stress relaxation, and pre-vulcanization characteristics (Mooney viscometer); ISO 89–1:2014: Rubber, unvulcanized – determinations using a shearing-disc viscometer – Part 1: Determination of Mooney viscosity.

Significance of Testing: These tests predict processing characteristics where a correlation between viscosity values and molecular mass exists. Viscoelastic characteristics can be correlated with stress relaxation. It can also predict the rate of cure and incipient cure time during very early stages of vulcanisation.

Procedure: A Mooney viscometer is the most common equipment used for these tests. The instrument consists of a motor-driven disk within a die cavity formed by two dies maintained at specified conditions of temperature and die closure force. Two rotors are used, a larger one with a diameter of 1.5 inch and a smaller one of 1.2 inch. The test determines the torque necessary to rotate the disk in a chamber filled with rubber. A number proportional to this torque is taken as an index of the viscosity of the rubber, where 8.3 ± 0.02 Nm equals 100 Mooney units (MU). The rotor is driven at 2 rpm. Generally, Mooney viscosity and stress relaxation tests are carried out at 100 °C and Mooney scorch at 135 °C. To test stress relaxation, after completion of the test time (usually 4 min), the test is run for another 2 min while the motor is at rest. The die and assembly of the Mooney viscometer rotor are shown in Figure 6.1.

Measurement and Calculation: Mooney viscosity M is usually expressed as 50 ML_{1+4} @ 100 °C, where 50 is the viscosity in MU, L indicates the large (standard) rotor, 1 is the preheating time, 4 is the test time and 100 °C is the test temperature. Other parameters that can be calculated by a Mooney viscometer include scorch time, cure index and optimum cure time. Typical Mooney cure curves are shown in Figure 6.2.

Limitations: The shear rates produced are not uniform but range from zero at the centre to a maximum at the periphery, and the average shear rate of this machine is ~2 sec^{-1} which is substantially below that encountered in extrusion and injection moulding. This test method cannot be used to study complete vulcanisation because the continuous rotation of the disk will result in slippage when the specimen reaches a stiff consistency.

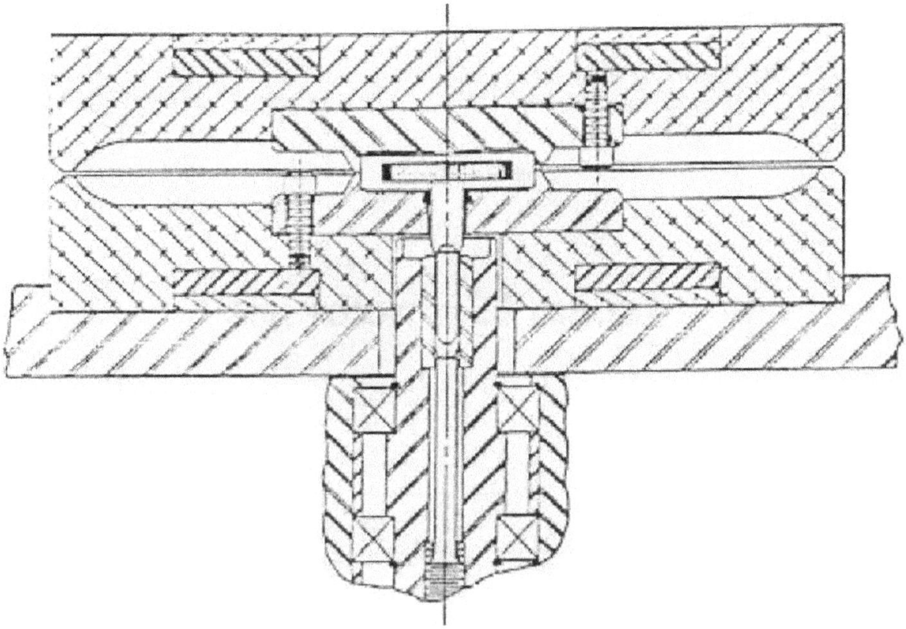

Figure 6.1: The die and assembly of a Mooney viscometer rotor. Reproduced with permission from *Rubber Science and Technology*, Ed., R. Mukhopadhyay, Shantinath Printing Udyog, Udaipur, Rajasthan, India, 2006. ©2006, Shantinath Printing Udyog [2]

6.3.5.3 Extrudability

ASTM D2230-96 (2012): Standard test method for rubber property – extrudability of unvulcanized compounds.

Significance of Testing: One objective in the extrusion of rubber compounds is to obtain a smooth extrusion that closely reproduces the contours of the extrusion die. This test method provides for a subjective determination of this under controlled conditions and it simulates to some extent the processing operations of extrusion and injection moulding. Apart from this it has the important advantages that much higher shear rates are possible than normally used in other instruments and it is comparable with those encountered in actual processing on the shop floor.

Procedure: In this test, rubber is forced through a small cylindrical die under a known pressure and the volume extruded in a given time is measured. Apart from the rate of extrusion it is also possible to visually estimate the quality of the extrusion. For this, the cylindrical die could be replaced by one with a more complicated cross-section (Garvey die) designed to show up typical faults that can occur in a mix with poor extrusion characteristics.

Measurement and Calculation: The Garvey die shown in Figure 6.3 has an acute angled wedge portion (X).

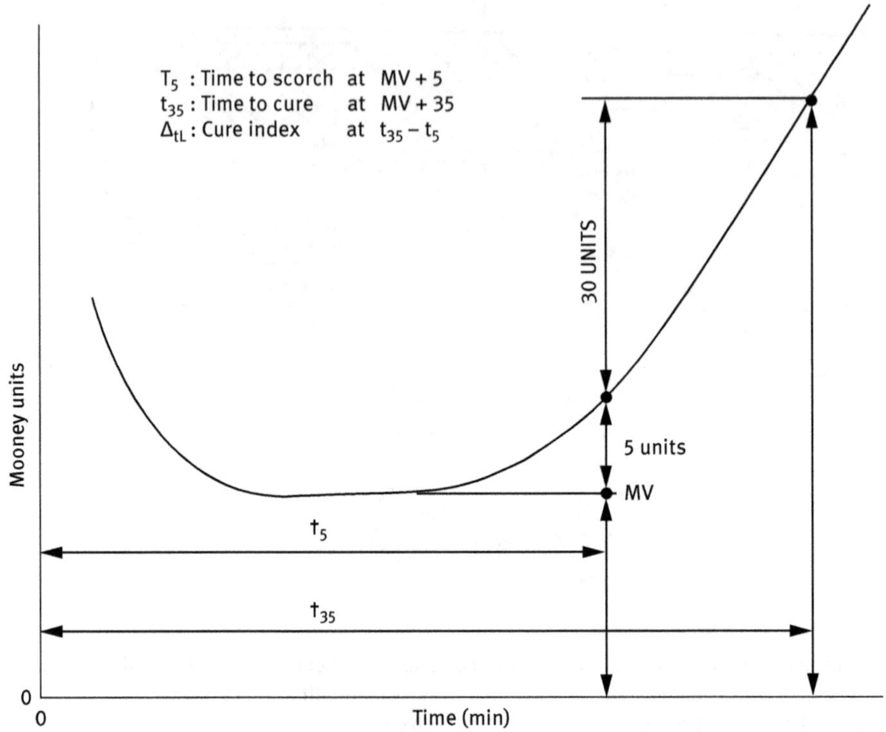

Figure 6.2: Typical Mooney cure curves. MV: minimum viscosity. Reproduced with permission from *Rubber Science and Technology*, Ed., R. Mukhopadhyay, Shantinath Printing Udyog, Udaipur, Rajasthan, India, 2006. ©2006, Shantinath Printing Udyog [2]

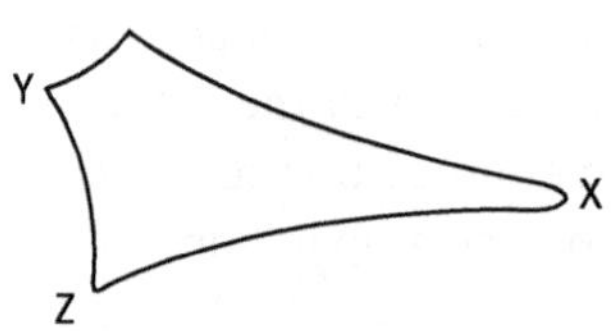

Figure 6.3: The Garvey die. Reproduced with permission from *Rubber Science and Technology*, Ed., R. Mukhopadhyay, Shantinath Printing Udyog, Udaipur, Rajasthan, India, 2006. ©2006, Shantinath Printing Udyog [2]

Mixes are rated by assigning a numerical value or score to each of the characteristics; for example, (i) surface; (ii) sharpness of the 'edge' X; (iii) the two 'corners' Y and Z; and (iv) cross-sectional dimensions (i.e., die swell). Rating figures can be taken from the above mentioned ASTM standard.

Limitations: As the rating is subjective, it does not lend itself readily to incorporate into a specification requirement. It does not measure other aspects of extrudability such as rate of extrusion or die swell in a quantitative manner. The short period of shearing is insufficient to break down thixotropic structures completely.

The vulcanisate tests are described in details (procedure, significance, standard used and so on) in Section 6.3.5.4.

6.3.5.4 Vulcanising characteristics

ASTM D2084-11: Standard test method for rubber property – vulcanization using oscillating disk curemeter; ASTM D5289-12: Standard test method for rubber property – vulcanization using rotorless curemeters; ISO 3417:2008: Rubber – measurement of vulcanization characteristics with the oscillating disc curemeter; and ISO 6502:1999: Rubber; measurement of vulcanization characteristics with rotorless curemeters.

Significance of Testing: This test is used to determine the vulcanisation characteristics of vulcanisable rubber compounds and is mainly used in the quality control process and R&D activity. The effect of compound changes on viscosity and scorch characteristics can be determined from the initial portion of the curve (shown in Figure 6.5), while the effect on the rate of the cure and the modulus of the cured compound can be detected in the latter portion of the curve. This is a very rapid test, so evaluation of the stock prior to final fabrication is easy. Good agreement between curemeter results and tensile modulus measurements make the curemeter an alternative method for controlling the production step and for testing raw materials like polymers and fillers. Even batch consistency can be checked easily with this equipment.

Procedure: In the oscillating disk-type rheometer (e.g., Monsanto R100) a biconical disk is embedded in the rubber in a closed cavity and the disk is oscillated through constant angular displacement (1, 3 and 5°). The curemeter assembly is shown Figure 6.4.

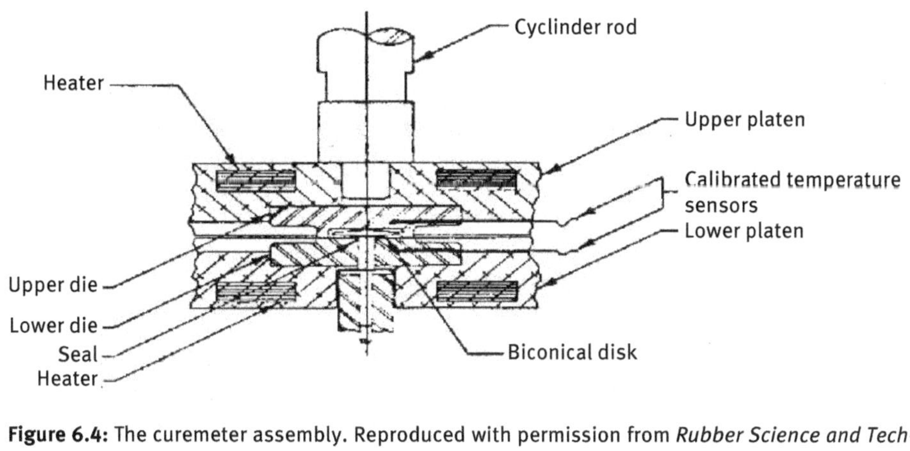

Figure 6.4: The curemeter assembly. Reproduced with permission from *Rubber Science and Technology*, Ed., R. Mukhopadhyay, Shantinath Printing Udyog, Udaipur, Rajasthan, India, 2006. ©2006, Shantinath Printing Udyog [2]

With a rotorless rheometer (e.g., Monsanto Moving Die Rheometer, MDR 2000), the sample is sealed within a cavity formed by the upper and lower dies, and oscillates through an arc of ±0.5, 1 and 3°. In both cases, torque is monitored and if needed can be plotted against time.

Measurement and Calculation: Typical curing curves obtained from an oscillating disk rheometer are shown Figure 6.5.

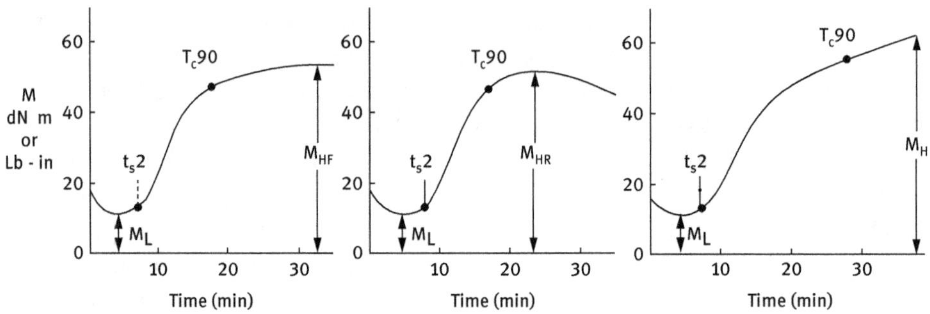

Figure 6.5: Typical curing curves obtained from an oscillating disk rheometer. Reproduced with permission from *Rubber Science and Technology*, Ed., R. Mukhopadhyay, Shantinath Printing Udyog, Udaipur, Rajasthan, India, 2006. ©2006, Shantinath Printing Udyog [2]

The information that can be deduced from a typical rheometer curve includes the minimum torque (M_L) and maximum torque (M_H), the M_{HF} [the maximum torque for a flat (plateau) curve], the scorch time (ts2, the time for two units rise in torque value from the M_L) and the optimum cure time (T_c) 90 [time corresponding to the optimum torque value that determines the time of moulding (minutes to M_L + 90 ($M_H - M_L$)/100 torque)].

Limitations of Oscillating Disk Rheometer over Moving Die Rheometer: The major advantages of the MDR over the oscillating disk rheometer are its high precision machine, its unique die design, a requirement for less sample, rapid thermal recovery, digital temperature control, an integral data system and so on.

6.4 Testing of rubber products

Tests on rubber products can be classified generally into two categories – compound tests and tests on actual products [16]. The former is mainly intended for compound evaluation, which is useful in quality control and developmental studies. Tests on finished retread tyre products are designed to give the performance characteristics of the retread product.

Rubber must be tested under carefully prescribed conditions. When products are made to a customer's satisfaction, it is essential that when the customer tests the product he is able to obtain the same results, within narrow limits, as the manufacturer obtained when the goods were produced. Similarly, raw materials are tested in the producer's laboratory and subsequently in the customer's laboratory, and the results must agree closely. For the retreading industry, the above conditions are also taken into consideration.

The international organisation that serves to standardise the test methods for rubber is the ASTM, which is supported by committees made up of industry representatives. In India the standardisation work is carried out by the Bureau of Indian Standards, known as ISI.

6.4.1 Preparation and conditioning of test pieces

Except in the case of complete products, a test piece must be formed and properly conditioned before the test can be carried out. Generally, the specimens are directly moulded or cut and/or buffed or sliced to some particular geometric shape. Specimen preparation and conditioning is done under a standard specification to yield more meaningful and repeatable test data. ASTM D3182-07 (2012) covers the same.

The properties of rubber depend on its history before test and the atmospheric conditions under which the test is carried out. In other words, the results are affected by the age of the rubber, the conditions under which it was stored (e.g., temperature and humidity), any mechanical deformation before test, and the humidity at the time of the test. Hence, to produce consistent results it is essential that these factors are controlled within appropriate limits. Virtually all test methods specify a conditioning period prior to test, in a standard atmosphere. The standard conditions are 23 ± 2 °C and 50 ± 5% relative humidity.

When both temperature and humidity are controlled, the standard conditioning time is a minimum of 16 h. Where only temperature is controlled, conditioning requires a minimum of 3 h.

6.4.2 Hardness

ASTM D2240-05 (2010): Standard test method for rubber property – durometer hardness; ASTM D1415-06 (2012): Standard test method for rubber property – international hardness; ISO 7619–1:2010: Rubber, vulcanized or thermoplastic – determination of indentation hardness; and ISO 48:2010: Rubber, vulcanized or thermoplastic – determination of hardness (hardness between 10 international rubber hardness degrees (IRHD) and 100 IRHD).

Significance of Testing: A hardness measurement is a simple way of obtaining the mechanical properties. This test is very common for any rubber product. The real purpose of running a hardness test is to determine the elastic modulus of a rubber under conditions of small strain by determining its resistance to a rigid indentor to which a force is applied. The hardness test can also predict the degree of cure. IRHD has a known relation to Young's modulus.

Procedure: In determining the hardness of a piece of rubber, the indentor is pressed down firmly under an applied force. The surface of the specimen must be flat enough for a uniform load to be applied. The sample must be ~8–10 mm in thickness and have sufficient surface area for the base of the indentor, or an incorrect value may be obtained.

Measurement and Calculation: In case of a Shore-A durometer, the pointer reads at its maximum in a scale of 0 to 100. The internationally accepted standard dead load hardness (IRHD) method covers rubbers in the range of 30–85 IRHD. In a normal test, a ball indentor with a diameter of either 2.38 or 2.5 mm is used, acting under a total force of 5.53 or 5.7 N, respectively. The indentation of the ball is made relative to the surface of the test piece after a fixed time of 30 sec after applying the force.

Limitations: Hardness results are criticised on two counts: (i) the measured values are unreliable because of mechanical considerations and the failure of operators to follow the standard technique of operation, and (ii) the characteristic that is measured, surface indentation, rarely bears any relation to the ability of a rubber product to function properly.

6.4.3 Tensile stress and strain including tear

ASTM D412-06a (2013): Standard test methods for vulcanized rubber and thermoplastic elastomers – tension; ASTM D624-00 (2012): Standard test method for tear strength of conventional vulcanized rubber and thermoplastic elastomers; ISO 1798:2008: Flexible cellular polymeric materials – Determination of tensile strength and elongation at break ISO 37:2005: Rubber, vulcanized or thermoplastic – Determination of tensile stress–strain properties and ISO 34–1:2010: Rubber, vulcanized or thermoplastic – Determination of tear strength – Part 1: Trouser, angle and crescent test pieces.

Significance of Testing: To the experienced rubber technologist, these tests are useful in many ways – for compound development, for manufacturing control and for determining a compound's susceptibility to deterioration by oil, heat, weathering and so on. These tests are almost universally used in tyre retreading industries as a means of determining the effect of various compounding ingredients and are particularly useful when such ingredients affect the rate and state of vulcanisation of the composition. The 'retention' properties are much more significant than the absolute values before and after exposure to deteriorating influences. The tear test can predict qualitatively the service performance of the product against any cut.

Procedure: The determination of tensile properties starts with a piece taken from the sample and includes the preparation and testing of the specimen. Specimens may be in the shape of a dumbbell, a ring or a straight piece of uniform cross-section.

Test pieces are strained at a constant rate of traverse, and force *versus* extension is recorded. The type of tensile dumbbell shown in Figure 6.6 and tear test specimens as in Figure 6.7 are used for this testing.

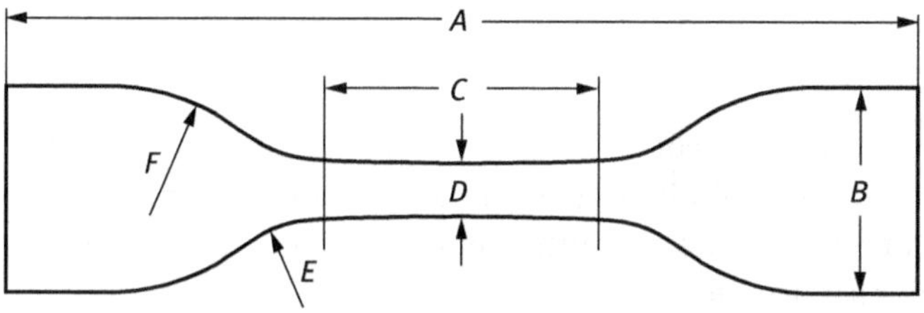

Dimensions	Type 1 (mm)	Type 2 (mm)
A	115 minimum	75 minimum
B	25 ± 1	12.5 ± 1
C	33 ± 2	25 ± 1
D	6.0 + 0.4 / 6.0 – 0.0	4 ± 0.1
E	14 ± 1	8 ± 0.5
F	25 ± 2	12.5 ± 1

Figure 6.6: Typical tensile dumbbell. Reproduced with permission from *Rubber Science and Technology*, Ed., R. Mukhopadhyay, Shantinath Printing Udyog, Udaipur, Rajasthan, India, 2006. ©2006, Shantinath Printing Udyog [2]

Measurement and Calculation: Tension tests are made on a power-driven machine able to produce a uniform rate of grip separation of 500 ± 50 mm/min. The force readings are expressed as stresses by reference to the original cross-sectional area of the test piece. The exact elongation is recorded by the extensometer and the load is monitored by the load cell attached to the universal testing machine. After completion of the test, modulus values at different elongation, tensile strength, and elongation at break can be determined or can be obtained directly from the computer if the instrument is equipped with a computer and proper software. Similarly, tear results can also be obtained after completion of the test.

Limitations: Tensile and tear tests have limited significance for the design or application engineer who is interested in determining the suitability of a composition

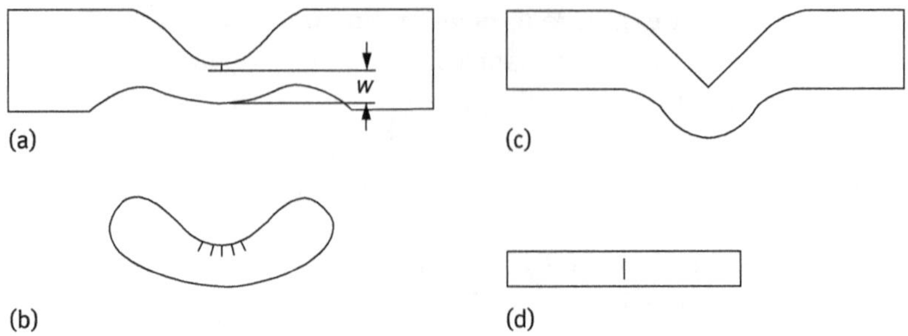

Figure 6.7: Typical tear specimen. Reproduced with permission from *Rubber Science and Technology*, Ed., R. Mukhopadhyay, Shantinath Printing Udyog, Udaipur, Rajasthan, India, 2006. ©2006, Shantinath Printing Udyog [2]

for use in a specific product. They cannot be used in design calculations and they bear little relation to a rubber part's ability to perform its function, because rubber products seldom approach their ultimate strength or elongation and are not subject to similar conditions in their service life.

6.4.4 Abrasion resistance

DIN 53516: Testing of rubber and elastomers; determination of abrasion resistance; ASTM D5963-04 (2010): Standard test method for rubber property – abrasion resistance (Rotary Drum Abrader); ASTM D1630-94 (2000): Standard test method for rubber property – abrasion resistance (Footwear Abrader); ASTM D2228-04 (2009): Standard test method for rubber property – abrasion resistance (Pico Abrader); and ISO 4649:2010: Rubber, vulcanized or thermoplastic – determination of abrasion resistance using a rotating cylindrical drum device.

Significance of Testing: Abrasion resistance is a performance factor of paramount importance for many rubber products, including tyres, retreads, conveyor belts, hoses, power transmission belts, footwear and floor coverings. For tyre and retread rubber products these tests give information on mileage. They may be used to estimate the relative abrasion resistance of different vulcanised rubber compounds and are suitable for comparative testing, quality control and R&D work.

Procedure: Various types of abrader are available, which use either loose or solid abradants. A loose abrasive powder can be used either in the manner of a shot blasting machine or between two sliding surfaces. Solid abradant consists of abrasive wheels, abrasive paper or cloth, and metal knives. The most commonly used instrument is the DIN Abrader, which uses a disk test piece (diameter of 16 mm, height of 6–10 mm) in a suitable holder under a load of 10 N, which is traversed across a rotating drum (diameter of 150 mm) with a rotating speed of 40 rpm covered with a sheet of emery paper and provides an abrasion path of 40 m. A running of a standard rubber sample is required to assess the abrasiveness of the emery paper.

Measurement and Calculation: Weight loss is measured after running the experimental sample in the DIN Abrader. Knowing the specific gravity of the sample and the standard abrasiveness of the emery paper, it is possible to calculate the volume loss of the experimental sample using Equation 6.3:

$$\text{Abrasion loss}(\text{mm}^3) = (\text{mass loss} \times 200)/(\text{specific gravity} \times \text{abrasiveness degree}) \tag{6.3}$$

Limitations: Because the conditions of abrasive wear in service are complex and vary widely, no direct correlation between these tests and actual performance can be assumed. The common mistake is to think of abrasion resistance as a single entity, like the modulus of elasticity. This is not the case. Modulus is a constant and remains the same, no matter in what product the compound is used. Abrasion resistance, on the other hand, is a complex of properties – resilience, stiffness, thermal stability, resistance to cutting and tearing – and different applications require these component properties in widely differing proportions. There are also some purely mechanical testing difficulties that may produce misleading results. For example, an undercured composition (soft and gummy) or one with a wax or oil 'bloom' will appear to have excellent abrasion resistance, merely because the abrasive has been clogged or lubricated by material removed from the surface of the specimen.

6.4.5 Resilience

ASTM D2632-14: Standard test method for rubber property – resilience by vertical rebound; ASTM D1054-91 (2000): Standard test method for rubber property – resilience using a rebound pendulum; and ISO 4662:2009: Rubber, vulcanized or thermoplastic – Determination of rebound resilience.

Significance of Testing: This is a very basic form of dynamic test and is a function of both the dynamic modulus and internal friction of a rubber. For the retreading industry, some knowledge of these dynamic properties is essential to understand the performance of products, which in service are subject to deformations of varying frequency and amplitude. These properties are important in rubber engineering applications such as springs and dampers.

Procedure: Rebound resilience can be measured by different types of equipment, including the Dunlop Tripsometer. This consists of a steel disk (diameter of 42 and thickness of 1.4 cm) mounted on bearings and with an out of mass balance in the form a bracket carrying the 4-mm-diameter indentor attached to its periphery. The scale is graduated in degrees to measure the angular displacement of the disk, which can be held in any desired position. The test piece should be a cylindrical block with an area of 8 mm_2 and thickness of 4 mm. The test specimen should be conditioned well before the test, especially if the testing is carried out at an elevated temperature. After loading the specimen, the disk is allowed to strike the specimen from an angle of 45° and the angular displacement of the rebound is noted.

Measurement and Calculation: Resilience is approximately related to the loss tangent by Equation 6.4:

$$R = E_R/E_T = \exp(-\prod \tan\delta) \quad (6.4)$$

or Equation 6.5:

$$E_A/E_R = \prod \tan\delta \quad (6.5)$$

where E_R is reflected energy, E_T is the incident energy, and E_A is the absorbed energy. With the Dunlop Tripsometer, rebound resilience is calculated as follows:

Rebound resilience (%) = (1 – cos α)/(1 – cos 45°), where α is the rebound angle and 45° is the initial angle.

Limitations: The magnitude of the resilience measurement depends on the size and shape of the pendulum hammer, the size, shape and stiffness of the rubber specimen, and the effective height of fall of the pendulum. The frequency is not of the same order of magnitude as that of many applications involving vibration.

6.4.6 Heat build-up

ASTM D623-07 (2014): Standard test methods for rubber property – heat generation and flexing fatigue in compression and ISO 4666–3:2010: Rubber, vulcanized – determination of temperature rise and resistance to fatigue in flexometer testing.

Significance of Testing: This test describes the heat rise in a vulcanised rubber cylinder after repeated compressive flexing and allows a fairly accurate prediction of the equilibrium operating temperature of rubber products such as tyres and retread products, V-belts and so on. During service, operating temperature will have an important effect on the rubber's physical properties. The data obtained with this test can be used to estimate the relative service quality of different compounds, and are often applicable to R&D studies.

Procedure: With the Goodrich flexometer, a definite compressive load (143 psi) is applied to a cylindrical test specimen (diameter of 17.8 ± 0.1 mm, height of 25 ± 0.15 mm) through a lever system with high inertia, while imposing on the specimen an additional high-frequency (1,800 ± 10 rpm) cyclic compression of definite amplitude (generally 4.45 mm). The performance of the machine is checked by studying a control compound.

Measurement and Calculation: The increase in temperature at the base of the test specimen is measured with a thermocouple after an interval or time (25 min) to provide a relative indication of the heat generated on flexing the specimen. By comparing the change in height, the 'permanent set' can be calculated, which indicates the degree of stiffening or softening of the test specimen. After completion of the test, ΔT is calculated, which is commonly known as the 'heat build-up'.

Limitations: Because of the wide variation in service conditions, no correlation between this accelerated test and service performance is given or implied. The results of flexometer tests are not suitable for control or classification purposes, nor are they applicable for use in purchase specifications.

6.4.7 Fatigue test

ASTM D4482-11: Test method for rubber property – extension cycling fatigue; ASTM D813-07 (2014): Standard test method for rubber deterioration – crack growth; and ISO 132:2005: Rubber, vulcanized – determination of flex cracking (De Mattia).

Significance of Testing: These tests are used to determine fatigue life at various extensions, compressions and repeated distortion by bending. These tests give an idea of the crack initiation behaviour of a rubber vulcanisate and only a very approximate measure of the crack propagation rate. It is possible to estimate the ability of a rubber vulcanisate to resist crack growth in a pierced specimen when subjected to flexing. The information obtained may be useful in predicting the flex-life performance of a compound in active service, such as in tyre tread and side wall compounds. For retread products this test gives information on the fatigue behaviour of the tread. These tests are useful in R&D work, especially when an unknown compound or construction is compared in a simultaneous test with a construction whose performance has been established in service.

Procedure: Different types of fatigue test are carried out, including the fatigue to failure test (FTFT), the De Mattia fatigue test, Ross fatigue, and so on. In each case the test specimens are cyclically strained at a fixed frequency and this cyclical straining action is called flexing. As a result of flexing, cracks, which are usually initiated by a naturally occurring flaw, grow and ultimately cause failure, which is defined as complete rupture of the test specimen. The number of cycles to failure (i.e., fatigue life) is recorded. Sometimes, an intentional cut is made on the specimen and the cut growth rate is measured after a certain interval. The test specimen for the De Mattia tester is usually a 6 × 1 inch moulded strip with a thickness of 6.4 ± 0.1 mm and a circular groove at the centre. For the FTFT, the specimen is a dumbbell with two beads at each end.

Measurement and Calculation: For the FTFT, six specimens are usually tested, and the four closest values are used to calculate the average fatigue life given in Equation 6.6:

$$\text{Japanese Industrial Standard average} = 0.5A + 0.3B + 0.1(C + D), \text{ where, } A > B > C > D \tag{6.6}$$

For the De Mattia test, three test specimens are generally tested. The data may be reported as the number of cycles required to reach a specified crack length, as the average rate of crack growth over the entire test period or as the rate of cracking in millimetres (or inches) per kilocycle during a portion of the test.

Limitations: No exact correlation between these test results and service is implied due to the varied nature of service conditions, and they are not suited for use in purchase specification requirements because the results from duplicate specimens do not agree with sufficient precision.

6.4.8 Ozone resistance

ASTM D1149-07 (2012): Standard test methods for rubber deterioration – cracking in an ozone controlled environment and ISO 1431–1:2004: Rubber, vulcanized or thermoplastic – Resistance to ozone cracking – Part 1: Static and dynamic strain testing, ISO 1431–2: Rubber, vulcanized or thermoplastic – Resistance to ozone cracking – Part 2: Dynamic strain test.

Significance of Testing: Ozone is a form of oxygen that occurs naturally in concentrations ranging from 2 to 20 parts per 100 million (pphm). Ozone attacks unsaturated rubbers like NR, SBR and acrylonitrile butadiene very rapidly, particularly when they are in a stressed condition. To address this problem, an accelerated ozone ageing test is carried out in a laboratory. This test is mainly used to differentiate, in a comparative sense, between different degrees of ozone resistance under limited and specified conditions. These tests basically give information about the protective power of the waxes and other antiozonants used in rubber vulcanisates.

Procedure: Different types of test specimens are used for different kinds of test. For static test, 6 × 1 inch rubber strips are given 20% stretch. For dynamic tests, 4 × 0.50 inch rubber strips are given 25% stretch. For bend loop tests, 3.75 × 1 inch rubber strips in a loop are subject to stretch varying from 0 to 25%. All specimens are preconditioned by being stretched for 24 h so that waxes and other antiozonants can come out at the surface. After pre-conditioning, the specimens are exposed to an ozone atmosphere of 50 pphm at a temperature of 40 °C for a defined period (~120 h).

Measurement and Calculation: After a certain interval the samples are checked with a magnifying glass and the first appearance of a crack is noted. Sometimes, a gradation is also given to identify the nature of the crack (0 means no crack and 4 means extensive cracks). It is always better to use a reference compound for which the values are known alongside the experimental one.

Limitations: Although these tests are helpful for comparative purpose, they may not correlate exactly with outdoor performance because service performance also depends on other conditions such as rainfall, ambient temperature, and so on. A factor that complicates the ozone cracking problem is the effect produced by exposure to sunlight, known is 'crazing', which is different from ozone cracks. Crazing occurs whether or not the specimen is elongated, and does not occur in a uniform direction but in all directions. This effect has been shown to be due to oxidation catalysed by sunlight and rainwater-soluble products. These cannot be assessed by indoor ozone testing.

6.4.9 Adhesion

ASTM D413-98 (2013): Standard test methods for rubber property – adhesion to flexible substrate.

Significance of Testing: These test methods determine the adhesion strength between plies of fabric bonded with rubber or the adhesion of the rubber layer in articles made from rubber attached to other material. These are applicable only when the adhered surfaces are approximately planar or uniformly circular, as in belting hose, tyre carcasses and so on.

These test methods are used to ensure the quality of a product by determining the force per unit width required to separate a rubber layer from a flexible substrate such as fabric, fibre or wire. For retread products adhesion is very important, as the tread has to adhere well to the casing.

Procedure: Strip specimen: Type A, 180° peel – measure the width of the strip cut to the nearest 0.22 mm (0.01 inch) and record. Separate the parts to be tested by hand at one end of the strip specimen and at a sufficient distance to permit the jaws of the testing clamp to be attached. Apply a specified or known mass by means of a clamp and mass carrier to the layer for which adhesion is being determined.

Measurement and Calculation: The adhesion value is reported as separation force per unit width.

Limitations: When the adhered surfaces contain sharp bends, angles or other gross irregularities that cannot be avoided in preparing test specimens, then special test methods must be used to evaluate adhesion. The acceptable value of adhesion measured will, of course, vary from product to product due to different rubber formulations, flexible substrates and according to type of product.

References

1. R.P. Brown in *Physical Testing of Rubber*, 4th Edition, Springer, New York, USA, 2006.
2. *Rubber Science and Technology*, Ed., R. Mukhopadhyay, Shantinath Printing Udyog, Udaipur, Rajasthan, India, 2006.

3. *Rubber Technology: Compounding and Testing for Performance*, Ed., J.S Dick, Hanser Gardner Publication, Cincinnati, OH, USA, 2001.
4. *Rubber Chemistry*, Ed., J.A. Brydson, Applied Science Publishers, London, UK, 1987.
5. *Rubber Technology and Manufacture*, Eds., C.M. Blow and C. Hepburn, Butterworth Publications, London, UK, 1982.
6. *Rubber Technology*, 2nd Edition, Ed., M. Morton, Van Nostrand Reinhold Company, New York, NY, USA, 1973.
7. *Rubber Products Manufacturing Technology*, Ed., A. Bhowmick, CRC Press, Marcel Dekker, Inc., New York, NY, USA, 1994.
8. *Basic Rubber Testing: Selecting Methods for a Rubber Test Program*, Ed., J.S Dick, ASTM International, Bridgeport, New Jersey, USA, 2003.
9. *The Science and Technology of Rubber*, 3rd Edition, Eds., J.E Mark, B. Erman and F.R. Eirich, Elsevier Academic Press, London, UK, 2005.
10. R.B. Simpson in *Rubber Basics*, Rapra Technology Ltd, Shawbury, Shrewsbury, UK, 2002.
11. M.J. Forrest in *Rubber Analysis: Polymers, Compounds and Products*, Rapra Technology Ltd, Shawbury, Shrewsbury, UK, 2001.
12. A. Ciesielski in *An Introduction to Rubber Technology*, Rapra Technology Ltd, Shawbury, Shrewsbury, UK, 1999.
13. *Current Topics in Elastomer Research*, Ed., A.K Bhowmick, CRC Press, Taylor & Francis, New York, NY, USA, 2008.
14. T.R. Crompton in *Polymer Reference Book*, Rapra Technology Ltd, Shawbury, Shrewsbury, UK, 2006.
15. *Annual Book of ASTM Standards*, Volume 9.01, ASTM International, Conshohocken, PA, USA, 2013.
16. *Annual Book of ASTM Standards*, Volume 9.02, ASTM International, Conshohocken, PA, USA, 2013.

Abbreviations

Accitard RE	*N*-cyclohexylthiophthalimide
AEG	American Engineering Group
API	American Petroleum Institute
ASTM	American Society for Testing and Materials
BET	Brunauer–Emmett–Teller
BR	Butadiene rubber
BSM	Benzthiazyl-sulphenmorpholide
CBS	Cyclohexyl benzthiazyl sulphenamide
CNT	Carbon nanotubes
CR	Chloroprene rubber
CSDPF	Carbon–silica dual-phase filler
CTAB	Cetyl-trimethyl-ammoniumbromide
CV	Constant viscosity
DBP	Dibutylphthalate
DDT	*N,N*-dimethyldodecylamine
DMA	Dynamic mechanical analysis
DPG	1,3-Diphenylguanidine
DPPD	*p*-Phenylenediamines
DSC	Differential scanning calorimetry
ENR	Epoxidised natural rubber
EPDM	Ethylene propylene diene monomer
ETRMA	European Tyre & Rubber Manufacturers' Association
FEF	Fast extrusion furnace
FTFT	Fatigue to failure rest
GPF	General purpose furnace
HAF	High abrasion furnace
HD	Highly dispersible
HMT	Hexamethylenetetramine
HSR	High-styrene resin(s)
IED	Improvised explosive devices
IIR	Isobutylene–isoprene rubber
IRHD	International rubber hardness degree
ISAF	Intermediate super abrasion furnace
ISO	International Organization for Standardization
ITRB	Industrial tint reference black
LDH	Layered double hydroxide
MBTS	Dibenzothiazyl disulfide
MC	Microcrystalline
MDR	Moving die rheometer
M_H	Maximum torque
M_{HF}	Maximum torque for flat (plateau) curve
M_L	Minimum torque
MMT	Montmorillonite
MT	Medium thermal
MU	Mooney unit(s)
MV	Minimum viscosity
MW	Molecular weight

https://doi.org/10.1515/9783110643893-007

MWD	Molecular macrostructures, degree of branching
NBR	Nitrile rubber
NDE	Non-destructive evaluation
NR	Natural rubber
NSA	Nitrogen surface area
OLS	Organic layered silicate
OMMT	Organo-modified montmorillonite
OTR	Off-the-road
PA	Process(ing) aid(s)
PBR	Polybutadiene rubber
PCN	Polymer clay nanocomposites
PPDA	Alkyl-aryl-*p*-phenylenediamines
PRI	Plasticity retention index
R&D	Research and development
RCN	Rubber clay nanocomposites
ROF	Run-on-flat
RSS	Ribbed smoked sheet
SAF	Super abrasion furnace
SBR	Styrene-butadiene rubber
SEM	Scanning electron microscopy
SMR	Standard Malaysian rubber
SP	Superior processing
SRF	Semi-reinforcing furnace
S-SBR	Solution-styrene butadiene rubber
Struktol HT 202	Blend of zinc soaps and fatty acid esters
STSA	Statistical thickness surface area
T_c	Optimum cure time
TEM	Transmission electron microscopy
TESPD	Bis(triethoxysilylpropyl)disulfane
TESPT	Bis(triethoxysilylpropyl)tetrasufide
$\mathrm{T_g}$	Glass transition temperature
TGA	Thermogravimetric analyser
TMQ	Trimethyl quinoline
TPMS	Tyre pressure monitoring system
t_s2	Scorch time
TSR	Technically specified rubber(s)
UV	Ultraviolet
UV-Vis	Ultraviolet–visible
WRAP	Waste & Resources Action Programme
XNBR	Carboxylated nitrile rubber

Index

https://doi.org/10.1515/9783110643893-008

www.ingramcontent.com/pod-product-compliance
Lightning Source LLC
Chambersburg PA
CBHW061210240926
51846CB00006B/1175